TRIPLE-NEGATIVE
BREAST
CANCER

TRIPLE-NEGATIVE
BREAST
CANCER

EDITORS

XIYUN DENG
Hunan Normal University School of Medicine, China

FAQING TANG
Hunan Cancer Hospital, China

THOMAS J. ROSOL
Ohio University Heritage College of Osteopathic Medicine, USA

World Scientific

NEW JERSEY · LONDON · SINGAPORE · BEIJING · SHANGHAI · HONG KONG · TAIPEI · CHENNAI · TOKYO

Published by

World Scientific Publishing Co. Pte. Ltd.

5 Toh Tuck Link, Singapore 596224

USA office: 27 Warren Street, Suite 401-402, Hackensack, NJ 07601

UK office: 57 Shelton Street, Covent Garden, London WC2H 9HE

Library of Congress Control Number: 2020940175

British Library Cataloguing-in-Publication Data
A catalogue record for this book is available from the British Library.

TRIPLE-NEGATIVE BREAST CANCER

ISBN 978-981-3277-75-5 (hardcover)
ISBN 978-981-3277-76-2 (ebook for institutions)
ISBN 978-981-3277-77-9 (ebook for individuals)

For any available supplementary material, please visit
https://www.worldscientific.com/worldscibooks/10.1142/11199#t=suppl

Typeset by Stallion Press
Email: enquiries@stallionpress.com

Contents

Foreword by Zhi-Ming Shao

Triple-negative breast cancer (TNBC) encompasses a subset of breast cancers that lacks significant expression of estrogen receptor (ER), progesterone receptor (PR), and human epidermal growth factor receptor 2 (HER2). TNBC has been increasingly recognized as a heterogeneous breast cancer subtype that displays characteristic features in terms of gene expression profiles, pathological features, as well as clinical behaviors. Since the introduction of chemotherapeutic-based adjuvant therapy for breast cancer decades ago, advances have been made in the treatment of TNBC. Recent developments in poly (ADP-ribose) polymerase inhibitors and immunotherapy have shed light on the management of a portion of TNBC cases. Despite this, precision medicine for individual TNBC patients by means of molecularly targeted therapy is still at its early stage and needs more research. Specifically, efforts based on understanding the biology of this specific breast cancer subtype, including heterogeneity and biomarkers, genetic and epigenetic characteristics, and strategies to enhance therapeutic responses are vital to the successful management of this deadly disease entity. In this regard, the effort to compile a book that is focused on TNBC is a well-timed contribution to the fields of scientific research and clinical practice. Hopefully, the researchers and the patients together with their families will benefit from this resourceful collection of scholarly works.

Zhi-Ming Shao, M.D., Ph.D.
Director, Department of Breast Surgery
Fudan University Shanghai Cancer Center
Director, Fudan University Cancer Institute
China

Foreword by Yibin Kang

Breast cancer is the most common female malignancy and the primary cause of cancer-related death among women worldwide. Management of breast cancer imposes a heavy economic burden on the healthcare system and the families involved. Although we have witnessed significant technological advances in treating breast cancer as a whole, a subset of breast cancers called triple-negative breast cancer (TNBC) still poses a major challenge to the care-providers and the sufferers. Lack of actionable therapeutic targets and early relapse are the major problems in the management of TNBC.

Since the first description of TNBC in the middle of last decade, it has acquired such a degree of scientific interest that, up to now, the term TNBC has appeared in more than ten thousand publications in the medical literature. This increase in the number of publications reflects the growing recognition of the importance of TNBC by the medical world. Nowadays, TNBC has become one of the most active fields in medical oncology research. This book is dedicated to providing up-to-date information to medical and research professionals, pharmaceutical companies, as well as TNBC patients and their supporters in a timely fashion. Although our perception of TNBC may change over time, from a professional's point of view, this book represents the first attempt to systematically describe the various aspects of TNBC. Future in-depth investigations on this unique disease entity will bring about significant improvement in the management of the disease and the quality of life of TNBC patients. Together, let's make TNBC history.

Yibin Kang, Ph.D.
Warner-Lambert/Parke-Davis Professor
of Molecular Biology
American Cancer Society Research Professor
Department of Molecular Biology
Princeton University
USA

About the Editors

Xiyun Deng is a professor of pathophysiology and molecular pathology and chair of Department of Basic Medical Sciences at Hunan Normal University School of Medicine located in Changsha, Hunan, China. He is the founder and currently serves as director of Key Laboratory of Translational Cancer Stem Cell Research at Hunan Normal University. Funded by the Natural Science Foundation of China (NSFC) and other funding agencies for over 15 years, his research interests cover the molecular mechanisms of cancer metastasis and, more recently, the experimental therapeutics targeting cancer stem cells. The major contributions he and his team have made over many years of research include: 1) the discovery that post-translational modifications of non-histone proteins play crucial roles in metastasis and drug-induced cellular stress responses in cancer stem cells; and 2) the development and mechanistic studies of cancer stem cell–targeting drugs against cancer with a particular focus on triple-negative breast cancer. He has published over 100 papers including first-author and/or corresponding-author articles/reviews in *Blood, Clinical Cancer Research, Cancer Treatment Reviews, ACS Applied Materials & Interfaces, International Journal of Cancer, European Journal of Cancer,* etc. As a research-oriented scholar, he is enthusiastically engaged in teaching activities. The undergraduates and graduates he mentored got several national-level awards including the "Challenge Cup" (2nd Prize), which is one of China's highest honors for undergraduates.

Faqing Tang is a professor of clinical pathology and laboratory medicine at Central South University and Jinan University, China. He serves as director of Clinical Laboratory and Medical Research Center, Hunan Cancer Hospital & the Affiliated Cancer Hospital of Xiangya School of Medicine, Central South University. He is director of Key Laboratory of Cancer Target Genes of Hunan Province, China. Dr. Tang has been engaged in various clinical trials including breast cancer. The research of the Tang laboratory focuses on the molecular carcinogenesis of cancer and the experimental therapeutics targeting oncogene. The major contributions he and his team have made include: 1) development and evaluation of new clinical laboratory technology, diagnostic reagents, and quality control software and approaches; 2) development of novel biomarkers for diagnosis/monitoring and therapeutic intervention of cancer including circulating tumor cells. He has published over 100 papers including first-author and/or corresponding-author articles in *Cancer Research, Journal of Biological Chemistry, PLoS One, BMC Cancer,* etc.

Thomas J. Rosol is a professor of veterinary and toxicologic pathology, Chair of Biomedical Sciences at the Ohio University Heritage College of Osteopathic Medicine and diplomate of the American College of Veterinary Pathologists. He has served as dean of the College of Veterinary Medicine and senior associate and interim vice president for research at Ohio State University and on advisory boards to the National Institutes of Health, United States Department of Agriculture, EPA, American Veterinary Medical Association, and Morris Animal Foundation. Rosol serves as a consultant for industry in preclinical safety and toxicology in the areas of endocrine, bone, and reproductive pathology and animal models of cancer. The Rosol laboratory investigates the pathogenesis of animal models of human cancer, mechanisms and treatment of bone metastasis, and has been funded by NIH for 30 years. The recent work focuses on breast cancer and other cancer types, and automated pathology using image analysis and artificial intelligence algorithms. Rosol has over 300 publications and is an elected fellow of the American Association for the Advancement of Science and was recognized by Ohio State University as a Distinguished Scholar, which is one of the university's highest honors.

List of Contributors

Ceshi Chen	Key Laboratory of Animal Models and Human Disease Mechanisms of the Chinese Academy of Sciences and Yunnan Province, Kunming Institute of Zoology, Chinese Academy of Sciences, Kunming, China
Sisi Chen	Key Laboratory of Translational Cancer Stem Cell Research, Hunan Normal University, Changsha, Hunan, China
Xiyun Deng	Key Laboratory of Translational Cancer Stem Cell Research, Hunan Normal University, Changsha, Hunan, China
Jingyang Du	Key Laboratory of Translational Cancer Stem Cell Research, Hunan Normal University, Changsha, Hunan, China
Junjiang Fu	Key Laboratory of Epigenetics and Oncology, Research Center for Preclinical Medicine, Southwest Medical University, Luzhou, Sichuan, China
Shujun Fu	Key Laboratory of Translational Cancer Stem Cell Research, Hunan Normal University, Changsha, Hunan, China
Xiaoxiang Guan	Department of Oncology, The First Affiliated Hospital of Nanjing Medical University, Nanjing, China
Guangchun He	Key Laboratory of Translational Cancer Stem Cell Research, Hunan Normal University, Changsha, Hunan, China
Doudou Huang	Department of Medical Oncology, Jinling Hospital, Medical School of Nanjing University, Nanjing, China

Md. Asaduzzaman Khan — Key Laboratory of Epigenetics and Oncology, Research Center for Preclinical Medicine, Southwest Medical University, Luzhou, Sichuan, China

Md. Shamsuddin Sultan Khan — EMAN Testing & Research Laboratory, Department of Pharmacology, School of Pharmaceutical Sciences, Universiti Sains Malaysia, Minden, Penang, Malaysia

Guifei Li — Key Laboratory of Translational Cancer Stem Cell Research, Hunan Normal University, Changsha, Hunan, China

Ying Li — Key Laboratory of Translational Cancer Stem Cell Research, Hunan Normal University, Changsha, Hunan, China

Meiling Liu — Key Laboratory of Chemical Biology and Traditional Chinese Medicine Research (Ministry of Education), College of Chemistry and Chemical Engineering, Hunan Normal University, Changsha, China

Lu Lu — Key Laboratory of Translational Cancer Stem Cell Research, Hunan Normal University, Changsha, Hunan, China

Qiujun Lu — Key Laboratory of Chemical Biology and Traditional Chinese Medicine Research (Ministry of Education), College of Chemistry and Chemical Engineering, Hunan Normal University, Changsha, China

Thomas J. Rosol — Department of Biomedical Sciences, Ohio University, Athens, Ohio, USA

Yuan Tan — Hunan Cancer Hospital & The Affiliated Cancer Hospital of Xiangya School of Medicine, Central South University, Changsha, Hunan, China

Faqing Tang   Hunan Cancer Hospital & The Affiliated
                   Cancer Hospital of Xiangya School of Medicine,
                   Central South University, Changsha, Hunan,
                   China

Mousumi Tania   Division of Molecular Cancer Biology, The
                   Red-Green Research Center, Dhaka,
                   Bangladesh

Cuiyan Wu   Key Laboratory of Chemical Biology and
                   Traditional Chinese Medicine Research
                   (Ministry of Education), College of Chemistry
                   and Chemical Engineering, Hunan Normal
                   University, Changsha, China

Mi Wu   Key Laboratory of Translational Cancer Stem
                   Cell Research, Hunan Normal University,
                   Changsha, Hunan, China

Yingying Wu   Key Laboratory of Animal Models and Human
                   Disease Mechanisms of the Chinese Academy
                   of Sciences and Yunnan Province, Kunming
                   Institute of Zoology, Chinese Academy of
                   Sciences, Kunming, China

Ting Xiao   Key Laboratory of Epigenetics and Oncology,
                   Research Center for Preclinical Medicine,
                   Southwest Medical University, Luzhou, Sichuan,
                   China

Shichao Yan   Department of General Surgery, The Second
                   Xiangya Hospital, Central South University,
                   Changsha, Hunan, China

Hui Yao   Key Laboratory of Translational Cancer Stem
                   Cell Research, Hunan Normal University,
                   Changsha, Hunan, China

Huimei Yi   Key Laboratory of Translational Cancer Stem
                   Cell Research, Hunan Normal University,
                   Changsha, Hunan, China

Liang Zeng — Department of Pathology, Guangzhou Women and Children's Medical Center, Guangzhou Medical University, Guangzhou, Guangdong, China

Qiuting Zhang — Key Laboratory of Translational Cancer Stem Cell Research, Hunan Normal University, Changsha, Hunan, China

Youyu Zhang — Key Laboratory of Chemical Biology and Traditional Chinese Medicine Research (Ministry of Education), College of Chemistry and Chemical Engineering, Hunan Normal University, Changsha, China

Chanjuan Zheng — Key Laboratory of Translational Cancer Stem Cell Research, Hunan Normal University, Changsha, Hunan, China

Ju Zhou — Key Laboratory of Epigenetics and Oncology, Research Center for Preclinical Medicine, Southwest Medical University, Luzhou, Sichuan, China

Abbreviations

ALDH	Aldehyde dehydrogenase
AMPK	AMP-activated protein kinase
AR	Androgen receptor
BCS	Breast conservative surgery
BCSC	Breast cancer stem cell
BLBC	Basal-like breast cancer
BLIS	Basal-like immune-suppressed
BRCA	Breast cancer susceptibility gene
ceRNA	Competitive endogenous RNA
ChoK	Choline kinase
circRNA	Circular RNA
CK	Cytokeratin
CNA	Copy number alteration
CSC	Cancer stem cell
CTC	Circulating tumor cell
ctDNA	Circulating tumor DNA
CTL	Cytotoxic T lymphocyte
CTLA4	Cytotoxic T lymphocyte-associated antigen 4
DFS	Disease-free survival
DMR	Differentially methylated region
DNMT	DNA methyltransferase
DNMTi	DNMT inhibitor
DSB	Double-strand break
EGFR	Epidermal growth factor receptor
EMT	Epithelial-to-mesenchymal transition
ER	Estrogen receptor
FISH	Fluorescence in situ hybridization

FZD	Frizzled
HAT	Histone acetyltransferase
HDAC	Histone deacetylase
HDACi	HDAC inhibitor
HER2	Human epidermal growth factor receptor 2
HIF	Hypoxia-inducible factor
HK2	Hexokinase 2
HR	Homologous recombination
HRD	Homologue recombination deficiency
HSP	Heat shock protein
IDC	Invasive ductal carcinoma
IHC	Immunohistochemistry
IM	Immunomodulatory
JAK	Janus kinase
LAR	Luminal androgen receptor
lncRNA	Long non-coding RNA
LRP	Lipoprotein receptor-related protein
MALAT1	Metastasis-associated lung adenocarcinoma transcript 1
MES	Mesenchymal-like
miRNA	MicroRNA
MRI	Magnetic resonance imaging
mTOR	Mammalian target of rapamycin
mTORC	Mammalian target of rapamycin complex
ncRNA	Non-coding RNA
NRTK	Non-receptor tyrosine kinase
OS	Overall survival
OXPHOS	Oxidative phosphorylation
PARP	Poly (ADP-ribose) polymerase
PARPi	Poly (ADP-ribose) polymerase inhibitor
PCho	Phosphocholine

PC-PLC	Phosphatidylcholine-specific phospholipase C
pCR	Pathologic complete response
PD1	Programmed cell death-1
PDK1	Pyruvate dehydrogenase kinase 1
PDL1	Programmed cell death-ligand 1
PFS	Progression-free survival
PI3K	Phosphatidylinositol-3 kinase
PIK3CA	PI3K p110 catalytic subunit
PKM2	M2 isoform of pyruvate kinase
PR	Progesterone receptor
PtdCho	Phosphatidylcholine
PTEN	Phosphatase and tensin homolog
RFS	Relapse-free survival
ROS	Reactive oxygen species
RTK	Receptor tyrosine kinase
SHH	Sonic hedgehog homologue
SSB	Single-strand break
STAT	Signal transducer and activator of transcription
TIL	Tumor-infiltrating lymphocyte
TK	Tyrosine kinase
TNBC	Triple-negative breast cancer
VEGF	Vascular endothelial growth factor
VEGFR	Vascular endothelial growth factor receptor

Chapter ONE
Overview of Triple-Negative Breast Cancer

Chanjuan Zheng[1,2], Guangchun He[1,2], *and* Xiyun Deng[1,2,*]

Contents

*Corresponding author: Xiyun Deng, E-mail: dengxiyunmed@hunnu.edu.cn
[1]Key Laboratory of Translational Cancer Stem Cell Research, Hunan Normal University, Changsha, Hunan, China.
[2]Departments of Pathology and Pathophysiology, Hunan Normal University School of Medicine, Changsha, Hunan, China.

1.1 Introduction

Breast cancer is by far the most common female cancer worldwide, with an estimated 2.09 million new cases diagnosed and 0.63 million deaths in 2018 globally [1]. These numbers represent a near 50% increase in incidence and 40% increase in mortality compared to ten years ago [2]. According to the International Agency for Research on Cancer (IARC), this trend of increase is expected to continue with an estimate of up to 3.06 million new cases diagnosed by the year 2040 [3]. In the United States, the American Cancer Society (ACS) estimates that more than a quarter of million women will be diagnosed with breast cancer each year, meaning approximately 1 out of every 8 US women is expected to develop breast cancer at some point in their lives [4].

Geographically and ethnically, significant differences in both incidence and mortality rates are documented. Notably, breast cancers are more common in developed countries than in developing countries. The highest incidence of breast cancer occurs in high-income regions such as North America, Northern/Western Europe, Australia, and New Zealand [4, 5]. Incidence is low throughout Africa, Asia, and most of Central and South America [6]. The risk factors behind breast cancer are complex but may involve environmental, lifestyle, and genetic factors, which collectively work together to determine cancer development.

Currently, several types of therapies are available to treat breast cancer, which include surgery, hormone therapy, chemotherapy, radiation therapy, and immunotherapy. Technological advancement continues to pave the way for improved therapies for breast cancer patients that adopt a targeted and personalized approach. However, one of the subtypes of breast cancer, called "triple-negative breast cancer (TNBC)", which is the topic of this book, represents a difficult-to-treat subtype of breast cancer, due to the lack of defined druggable molecular targets.

1.2 Classification of Breast Cancer

1.2.1 *Histopathological Classification*

Histomorphologically, at least 18 different types of breast cancer are described by the World Health Organization. The histopathological types of invasive breast cancer that are commonly encountered by pathologists are summarized in **Table 1-1**. Among these different types,

Table 1-1. Histopathological types of invasive breast cancer.

Histopathological type	Frequency	10-year survival rate
Invasive ductal carcinoma not otherwise specified (IDC NOS)	50–80%	35–50%
Invasive lobular carcinoma (ILC)	5–15%	35–50%
Mixed type, lobular and ductal features	4–5%	35–50%
Medullary carcinoma	1–7%	50–90%
Tubular/invasive cribriform carcinoma	1–6%	90–100%
Neuroendocrine carcinoma	2–5%	Unknown
Metaplastic carcinoma	<5%	Unknown
Invasive micropillary carcinoma	<3%	Unknown
Mucinous carcinoma	<5%	80–100%
Invasive apocrine carcinoma	0.3–4%	35–50%
Adenoid cystic carcinoma	0.1%	Unknown

Ref: *Nat Rev Cancer. 2005; 5(8):591–602.*

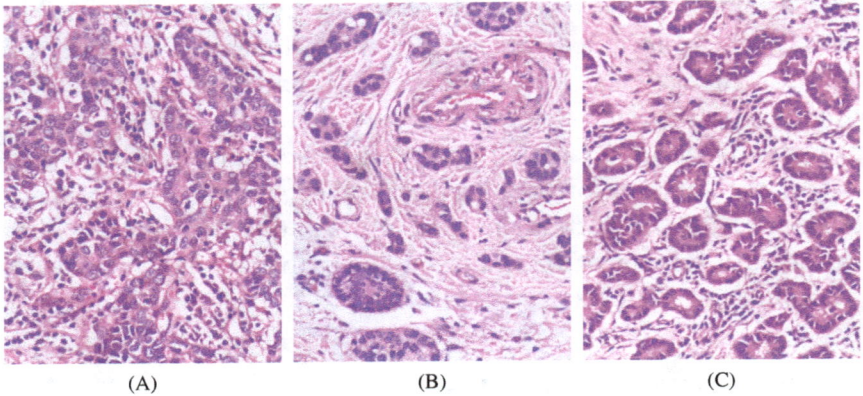

(A)   (B)   (C)

Figure 1-1. Histomorphological images of common types of breast cancer. (A) Invasive ductal carcinoma; (B) Invasive lobular carcinoma; (C) Tubular carcinoma. Images (H&E-stained, original magnification at 200×) courtesy of Dr. Songqing Fan, Department of Pathology, The Second Xiangya Hospital, Central South University, Hunan, China.

invasive ductal carcinoma (IDC) not otherwise specified (IDC NOS) is by far the most commonly encountered type, accounting for 50–80% of all breast cancers. Invasive lobular carcinoma is the second most common type of breast cancer, followed by other less commonly encountered breast cancer types such as the mixed lobular and ductal type, medullary carcinoma, and tubular carcinoma, etc. [7] (**Figure 1-1**).

1.2.2 *Molecular Classification*

1.2.2.1 *Molecular classification based on gene expression profiling*

Although histopathological classification has played some roles in defining tumor characteristics, this way of categorizing breast cancers fails to separate the tumors into different entities with type-specific prognosis and treatment options. In addition, histopathological classification may not be accurate since it depends largely on the pathologist [8]. Since the beginning of this century, Dr. Perou and colleagues at Lineberger Comprehensive Cancer Center, University of North Carolina at Chapel Hill adopted a new molecular classification approach based on gene expression profiling to classify breast cancers into "intrinsic" subtypes [9, 10].

Gene expression profiling based on cDNA microarray revealed that the gene expression differs enormously between tumors that are positive for estrogen receptor (ER) and progesterone receptor (PR) and tumors that are negative for these hormone receptors. ER/PR-positive and ER/PR-negative expression is characteristic of luminal and basal cells, respectively. Correspondingly, ER/PR-positive tumors are clustered into two subgroups with expression patterns similar to luminal epithelial mammary cells, i.e., luminal A and luminal B. ER/PR-negative tumors are clustered into three distinct molecular subgroups, including tumors with gene expression similar to basal/myoepithelial mammary cells, i.e., basal-like; tumors with characteristics of gene amplification of human epidermal growth factor receptor 2 (HER2), i.e., HER2-enriched; and tumors with expression patterns related to normal mammary stromal cells, i.e., normal-like. Later on, it was discovered that the normal-like breast carcinomas do not seem to constitute a true subtype and may be due to contamination of normal breast cells. Subsequently, through additional gene expression analysis, the same group of researchers from University of North Carolina at Chapel Hill showed that basal-like breast tumors included another subtype, termed claudin-low [11]. Both basal-like and claudin-low breast cancers are predominantly (but not exclusively) ER-, PR-, and HER2-, so called "TNBC" for not expressing any of these receptors. Therefore, the intrinsic subtypes based upon gene expression profiling include a total of five subtypes, which can be grouped into ER+ (luminal A and luminal B), HER2+ (HER2-enriched), and triple-negative (basal-like and claudin-low) [12].

Whether or not these intrinsic subtypes reflect different cells of origin or different differentiation pathways is a subject of some debate. However, they clearly help to explain the differences in the behaviors of breast tumors and their responses to treatment despite apparent morphological similarity. Notably, these different subtypes are indeed associated with distinct prognosis and treatment options. While the luminal subtypes are associated with satisfactory outcomes,

TNBC patients have a poorer overall survival (OS) compared with other subtypes.

1.2.2.2 *Molecular classification based on immunohistochemistry*

Although gene expression profiling has been regarded as the gold standard for molecular classification of breast cancer, its use in clinical practice has been limited by strict tissue requirements and by issues of cost, technical complexity, and potential batch effect of gene expression profiling [13]. In an attempt to develop a molecular classification method that is clinically significant, technically simple, reproducible, and readily available, investigators have used immunohistochemistry as a surrogate for cDNA microarray in performing molecular classification of breast cancer. This immunohistochemistry-based molecular classification has proven able to provide diagnostic, prognostic, and predictive information in breast cancer and is thus widely accepted by pathologists and clinicians.

For practical immunohistochemistry-based classification, in addition to the typical three receptors, i.e., ER, PR, and HER2, immunostaining for cytokeratin (CK) and epidermal growth factor receptor (EGFR [HER1]) is also performed. Therefore, subtypes of breast cancer based upon immunohistochemical marker analysis include (**Table 1-2**): luminal A (ER+ and/or PR+, HER2-), luminal B (ER+ and/or PR+, HER2+), triple-negative or basal-like (ER-, PR-, HER2-, EGFR (HER1)+ and/or CK5/6+), and HER2-enriched (ER-, PR-, HER2+) [14].

While molecular classification by gene expression profiling is burdensome and costly and, therefore, cannot be used on a daily basis,

Table 1-2. Subtypes of breast cancer based on immunohistochemical markers.

Molecular subtype	Immunohistochemical profile	Frequency
Luminal A	ER+ and/or PR+, HER2-	50–60%
Luminal B	ER+ and/or PR+, HER2+ or HER2-/Ki67 ≥14%	10–20%
HER2-enriched	ER-, PR-, HER2+	10–15%
Triple-negative or Basal-like	ER-, PR-, HER2-, EGFR+ and/or CK5/6+	10–20%

Refs: *Breast Cancer Res Treat. 2008; 109(1):123–139; Cancer Treat Rev. 2012; 38(6):698–707.*

immunohistochemistry is routinely performed in the pathology lab for molecular classification of breast cancer instead.

1.3 Definition and Diagnosis of TNBC

By definition, TNBC is characterized by the lack of clinically significant levels of ER and PR as well as HER2 amplification or overexpression [15]. The levels of ER, PR, and HER2 proteins are assessed by immunohistochemistry. In addition, gene amplification of HER2/neu is assessed by fluorescence in situ hybridization (FISH), with or without immunohistochemistry. According to the MD Anderson Cancer Center pathological diagnosis criteria for TNBC, the status of ER and PR is determined using immunohistochemistry, with a cutoff of less than 10% for negativity, and HER2/neu status is considered negative either immunohistochemistry is 0 to 1+ without FISH or FISH result is negative. The diagnosis criteria of TNBC are summarized in **Table 1-3**.

A typical immunohistochemical staining pattern for TNBC and receptor-positive breast cancer is illustrated in **Figure 1-2**.

Later on, the American Society of Clinical Oncology (ASCO) and the College of American Pathologists (CAP) have recommended a change of the cutoff for the two hormone receptors (ER and PR) from 10% to 1% [16]. This lower cutoff is based on the observation that there is a benefit from endocrine therapy, even for breast tumors with very low level of hormone receptors. Accordingly, the number of patients with TNBC will be greatly decreased, if this recommendation is widely accepted by pathologists and clinicians into routine practice. More details about the diagnosis of TNBC by immunohistochemistry and/or FISH will be discussed in Chapter 6.

Table 1-3. Diagnosis criteria of TNBC.

Protein name	Gene	Detection method (cutoff)
ER (Estrogen receptor)	ESR1	Immunohistochemistry (<10%)
PR (Progesterone receptor)	PGR	Immunohistochemistry (<10%)
HER2 (Human epidermal growth factor receptor 2)	ErbB2/neu	Immunohistochemistry (0–1+) or FISH (–)

Ref: *J Cancer. 2017; 8(11):2026–2032.*

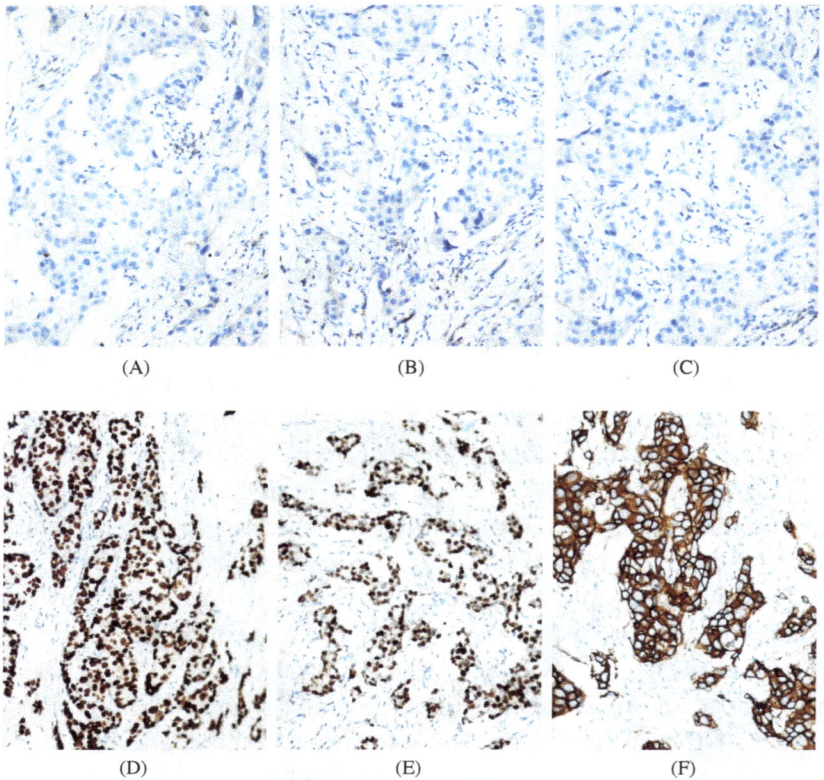

Figure 1-2. Immunohistochemical staining of TNBC vs. receptor-positive breast cancer. (A, B, C) Typical negative immunohistochemical staining for ER (A), PR (B), or HER2 (C) in TNBC. (D, E, F) Positive immunohistochemical staining for ER (D), PR (E), or HER2 (F). The nuclei are counterstained with hematoxylin (blue color) to reveal the tissue architecture. Images (original magnification at 200×) courtesy of Dr. Liang Zeng, Department of Pathology, Hunan Cancer Hospital & The Affiliated Cancer Hospital of Xiangya School of Medicine, Central South University, Hunan, China.

1.4 Similarities and Differences Between TNBC, Basal-Like Breast Cancer, and BRCA-Associated Breast Cancer

1.4.1 *TNBC vs. Basal-Like Breast Cancer (BLBC)*

The term "basal-like" is derived from the similarity of the gene expression signature of this molecularly defined subtype with that of the normal basal myoepithelial cells of the breast. As the name

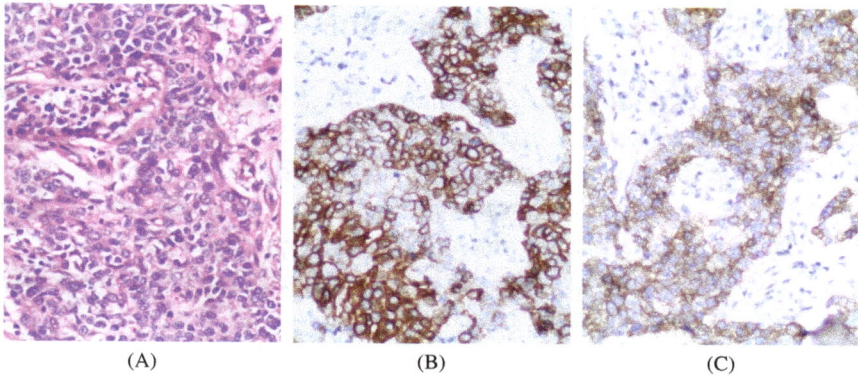

(A)   (B)   (C)

Figure 1-3. Histopathological and immunohistochemical features of basal-like breast cancer. Histopathological features of basal-like breast cancer (A) together with positive immunohistochemical staining of cytokeratin 5/6 (B) and EGFR (C) are shown. Images (H&E-stained for A and hematoxylin-counterstained for B and C; original magnification at 200×) courtesy of Dr. Songqing Fan, Department of Pathology, The Second Xiangya Hospital, Central South University, Hunan, China.

suggests, basal-like breast cancer (BLBC) expresses genes that are expressed in cells of the normal, non-luminal (basal) myoepithelial layer of the mammary ducts and lobules. The basal-like or basal-type genes which are normally expressed include those that are important for structural integrity and for cell-matrix interactions [17]. According to Lineberger Comprehensive Cancer Center, University of North Carolina at Chapel Hill, BLBC is defined by immunohistochemistry as ER−, PR−, HER2−, EGFR+ and/or CK5/6+ (**Figure 1-3**) [14].

Since BLBC is associated with the triple-negative phenotype, TNBC has initially been perceived as a synonym with BLBC. Many studies have used the absence of the three receptors, sometimes along with positive expression of CK5/6 (or CK17) and/or EGFR, as a characteristic feature to define BLBC in immunohistochemical staining. In the literature, BLBC and TNBC are frequently used interchangeably. It should be noted, however, although there is significant overlap between TNBC and BLBC, these terms are not synonymous. Indeed, a variety of studies have demonstrated that TNBC displays a great deal of heterogeneity and that these two definitions are not synonymous. The following facts may help to clarify the similarities and differences between TNBCs and BLBCs.

Similarities between TNBC and BLBC:

(1) *Most TNBCs fall into the category of the basal-like subtype as demonstrated by the fact that approximately 80% of TNBC overlaps with the BL phenotype at the transcriptomic level.*

(2) *BLBCs share many features that are associated with TNBC, such as high histological grade, elevated mitotic count, and expression of EMT markers.*

(3) *BRCA1-mutated breast cancers (described below) demonstrate characteristics of both TNBC and BLBC.*

Differences between TNBC and BLBC:

(1) *While most BLBCs are triple-negative, about 1/3 of BLBCs are not; conversely, although the majority of TNBCs express basal markers at the protein level, a significant proportion (around 10–35%) of TNBCs are not basal-like.*

(2) *While testing for TNBC has become quite routine in clinical practice, the identification of the basal-like status remains burdensome due to the requirement of cDNA microarray analysis.*

Although TNBC does not form a homogeneous group of disease when analyzed by gene expression profiling, it is believed that the basal-like subtype does form a homogeneous group of tumors with a similar gene expression profile related to prognosis and therapy response [18, 19]. This indicates that the poor prognosis of TNBC may actually reflect the high percentage of triple-negative tumors that are of basal-like. Indeed, a survival analysis of over 900 cases of breast cancer illustrated a shorter disease-specific survival among cases that expressed the basal markers CK5/6 and 17. In addition, expression of EGFR (HER1) has been found to be an independent negative prognostic factor for TNBC patients (relative risk [RR] 1.54) [20].

1.4.2 TNBC vs. BRCA-Associated Breast Cancer

Breast cancer susceptibility genes (BRCAs) are tumor-suppressor genes, carriers of whose mutations are at an increased risk of developing cancer in the breast, ovary, and others [21]. While BRCA2 function is limited to DNA recombination and repair processes, BRCA1 seems to have relatively broader cellular functions, which involve DNA repair and gene transcription regulation [22]. The functions of BRCA genes in TNBC will be discussed in more detail in Chapter 3 and Chapter 5.

In addition to its close relationship with BLBC, TNBC is also highly associated with the BRCA1 status. The following are several lines of evidence demonstrating the close relationship between BRCA abnormalities and the TNBC phenotype.

Relationship between TNBC and BRCA abnormalities:

(1) *The majority of BRCA1/2-mutated breast cancers exhibit a triple-negative phenotype [23].*
(2) *Up to 80% of hormone receptor-negative breast cancers have reduced or undetectable BRCA1 expression [24].*
(3) *Although germline mutations in BRCA1/2 are low in TNBC patients (10–20%) [25], these mutations can confer a lifetime risk of up to 85% of developing breast cancer, with 90% of these being triple-negative [26].*
(4) *BRCA1-associated breast tumors share biological similarities with TNBC and/or BLBC, including younger age at diagnosis and high tumor grade.*

Cancers that lack functional BRCA1 or BRCA2 have a deficiency in the repair of DNA double-strand breaks (DSBs) by a process called homologous recombination. This deficiency results in genomic instability and susceptibility to drugs that generate DNA strand breaks, which include alkylating agents (e.g., platinum and mitomycin C) and PARP inhibitors (PARPi's) (e.g., olaparib and talazoparib). In cells with homologous recombination deficiency, these drugs cause persistent DNA damage and consequently, cell death, a phenomenon called "synthetic lethality", which will be discussed in more detail in Chapter 6.

1.5 Epidemiology and Risk Factors of TNBC

Although some inconsistencies have been reported in the literature, generally speaking, TNBC accounts for 10–20% of all breast cancer cases. Basically, not much epidemiological risk assessment has been performed in large population-based studies on the contribution of race and other factors to TNBC and/or BLBC. Even so, it is clear from the literature that the distribution of TNBC/BLBC differs significantly by race and menopausal status. Age, sex, and other factors also contribute to the occurrence of TNBC among all population groups.

1.5.1 *Distribution of TNBC/BLBC by Race and Menopausal Status*

It has been noted by many research groups that the frequency of TNBC/BLBC is consistently higher in populations of women of African ancestry than in other racial or ethnic groups at all ages. An analysis of the US Surveillance, Epidemiology, and End Results (SEER) data of women diagnosed with breast cancer in 2010 provided evidence to show that African American (odds ratio [OR] 1.40 [1.20–1.60]) and Hispanic women (OR 1.30 [1.20–1.50]) were more likely to be diagnosed with TNBC than white women. Studies from the US-based cancer databases show that among the different races or ethnic groups, African American (non-Hispanic black) women have the highest incidence of TNBC followed by Hispanic women. Non-Hispanic whites and Asian women in the US have similar low percentage of TNBC relative to all breast cancer cases [27, 28] (**Figure 1-4**). Similar results are reported from a retrospective cohort study of patients with breast cancer in the UK, showing that 22% of black women have TNBC compared with 15% of white women who have TNBC [29].

Reports from Africa further confirmed the high frequency of TNBC in women of African ancestry. A study in Nigeria and Senegal (507 women, mean age 44.8 years) showed that TNBC, including basal-like TNBC, was the predominant type of invasive breast cancer (27%) [30]. Of consecutive cases of breast cancer reported in a Bamako University hospital in Mali, the mean age of

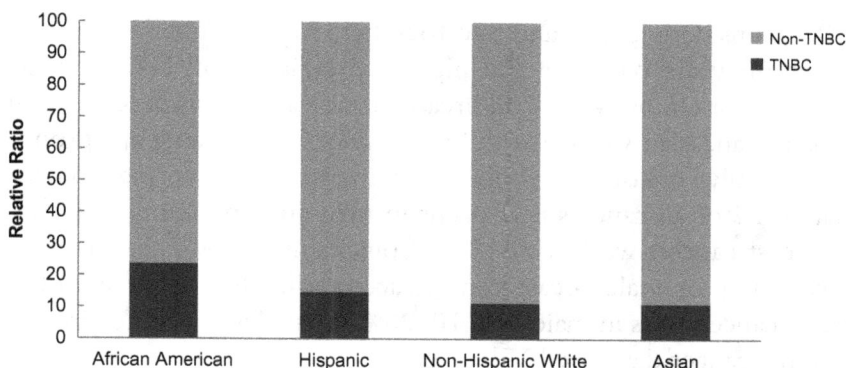

Figure 1-4. Relative rate of TNBC in different ethnic groups in the United States. Among the different races or ethnic groups in the USA, non-Hispanic black women have the highest incidence of TNBC followed by Hispanic and non-Hispanic white women. Women of Asian and other ethnic groups have similar incidence of TNBC as non-Hispanic whites.

patients was 46 years (range 25–85 years) and 46% of tumors were triple-negative.

Menopausal status has also been considered as a composite factor together with race. In a population-based North Carolina Breast Cancer cohort of 878 African American women with breast cancer, premenopausal African American women had higher rates of BLBC (39%) than did white women of similar age (16%) or postmenopausal African American women (14%) [31]. Collectively, TNBCs and/or BLBCs are more common in premenopausal women of African ancestry.

1.5.2 *Distribution of TNBC/BLBC by Age and Sex*

In addition to the above-mentioned race and menopausal status, TNBCs have a tendency to occur in younger women than their non-TNBC counterparts. Data from the California Cancer Registry revealed that the odds of a woman having TNBC under the age of 40 years was 1.53 compared with a woman at an age greater than 40

[28]. Considering race and age together, African American women under 50 years old have the highest prevalence of TNBC, which accounts for almost 40% of all breast cancer cases, compared to about 15% in Caucasian women with breast cancer in the same age group.

Basically, breast cancer in males is rare, representing approximately 1% of all cancers that occur in men and approximately 1% of all breast cancers worldwide [32]. Coincidently, TNBC is present in fewer cases of male breast cancer, accounting for about 6% of all breast cancer cases in males (vs. 10–20% in females) examined in a US national cancer database [27].

1.5.3 *Other Risk Factors of TNBC/BLBC*

Other risk factors have also been identified for TNBC. Millikan and colleagues conducted a population-based case-control study that involved approximately 1,400 women with invasive breast cancer [14]. The authors noted that compared with women with luminal A tumors, those with basal-like tumors were more likely to have increased parity and younger age at full-term pregnancy. Furthermore, unlike women with luminal A tumors, basal-like tumors were more likely in women with younger age at menarche, who breast fed for shorter durations and had a higher body mass index. It is hoped that programs aimed at promoting breast-feeding and reducing abdominal adiposity would reduce the number of cases of TNBC/BLBC among women, particularly younger African American women.

In addition, low socioeconomic status is associated with many of the shared characteristics of breast tumors that occur in women of African ancestry, including high grade, high clinical stage, and ER-negative status. Data from the California Cancer Registry showed that irrespective of race or ethnic origin, women living in areas of low socioeconomic status are more likely to be diagnosed with TNBC than women living in areas of high socioeconomic status [28]. This observation is in contrast with the association of all-type breast cancers with high (rather than low) socioeconomic status, discussed above (under Introduction).

Table 1-4. Risk factors associated with TNBC/BLBC.

Women of African ancestry

Premenopausal status

Younger age at diagnosis (<40 years)

High parity

Younger age at first full-term pregnancy

Younger age at menarche

Shorter duration of breast feeding

Elevated body mass index

Lower socioeconomic status

Refs: *Clin Breast Cancer. 2009; 9(Suppl 2):S73–S81; Semin Oncol. 2011; 38(2):254–62; Breast Cancer Res Treat. 2008; 109(1):123–39; Lancet Oncol. 2014; 15(13):e625–e634.*

The risk factors of TNBC/BLBC are summarized in the table above (**Table 1-4**).

1.6 Characteristics of TNBC

1.6.1 *General Clinical Features of TNBC*

While TNBCs constitute 10–20% of breast cancer incidence, they account for almost half of all breast cancer deaths. As discussed above, TNBCs are more prevalent in women of African ancestry and more frequently strike younger patients. TNBC tumors are generally larger in size, are of high grade and less differentiated, and are biologically more aggressive. TNBC patients have a higher rate of distant recurrence and a poorer prognosis than those with other subtypes. Less than 30% of women with metastatic TNBC survive 5 years and almost all die of their disease despite adjuvant chemotherapy [33].

Compared with other subtypes of breast cancer, TNBC tumors are 2.5 times more likely to metastasize within 5 years of diagnosis. TNBC preferentially metastasizes to the viscera (including the lungs, brain, and liver) in contrast to non-TNBC which disseminates mostly to the bone. Resultantly, median time to death is shorter and OS is poorer for patients with TNBC compared with those with non-TNBC.

Because of the lack of receptor targets existing in other subtypes of breast cancer, TNBC patients do not benefit from endocrine or anti-HER2 therapy [34]. Although significant advances have been made in identifying potential targets that can be used for the development of novel therapeutic strategies (discussed in Chapter 7 and Chapter 8), chemotherapy remains the only established systemic therapeutic option for these patients. In spite of initial responses to chemotherapy, drug resistance develops rapidly and the prognosis of metastatic TNBC is poor. More details about the general clinical features of TNBC will be discussed in Chapter 6.

1.6.2 *Immunophenotypic and Molecular Characteristics of TNBC*

As discussed above, besides lacking hormone receptors and HER2, TNBCs are more likely to express myoepithelial markers, such as CK5/6, c-Kit, and less likely to express epithelial markers, such as E-cadherin. Since HER2 is negative in TNBC, EGFR (HER1), the other member of receptor tyrosine kinase family proteins, is usually overexpressed. More than half of TNBCs have protein abnormality or gene mutation of p53. The PI3K pathway is commonly activated in TNBC, although the mutation rate of PI3K is low. The immunophenotypic and molecular characteristics of TNBC/BLBC will be described in more detail in Chapter 6.

1.6.3 *Histopathological Characteristics of TNBC*

TNBC patients usually have larger tumors than hormone receptor-positive patients and are more likely to be of high grade, have lymphovascular invasion, and present with clinically metastatic disease. Adjusting for tumor size and grade in a multivariable logistic regression model, TNBCs are found to have a much lower rate of lymph node positivity than any other subtype (OR 0.59 [0.57–0.61]). Histopathologically, the vast majority (>90%) of TNBC tumors belong to IDC; other types such as invasive lobular carcinoma, metaplastic carcinoma with squamous differentiation, spindle-cell metaplastic carcinoma, adenoid cystic carcinoma, and secretory carcinoma are occasionally seen in TNBC [35]. These non-IDC-type TNBCs may have a better prognosis than the usually poor-prognosis TNBC in general.

TNBCs also have been reported to have a high mitotic index, high Ki67 expression, central necrosis, pushing margins, and dense lymphocytic infiltrate [17, 20]. In addition, high degree of aneuploidy and nuclear pleomorphism are also regarded as characteristics of these tumors. More details about the histopathological characteristics of TNBC will be discussed in Chapter 6.

References

1. Bray F, Ferlay J, Soerjomataram I, Siegel RL, Torre LA, Jemal A. (2018). Global cancer statistics 2018: GLOBOCAN estimates of incidence and mortality worldwide for 36 cancers in 185 countries. *CA Cancer J Clin*, **68**(6): 394–424.

2. Vinayak S, Schwartz EJ, Jensen K, Lipson J, Alli E, McPherson L, Fernandez AM, Sharma VB, Staton A, Mills MA *et al.* (2013). A clinical trial of lovastatin for modification of biomarkers associated with breast cancer risk. *Breast Cancer Res Treat*, **142**(2): 389–398.

3. Hashemi SM, Balouchi A, Al-Mawali A, Rafiemanesh H, Rezaie-Keikhaie K, Bouya S, Dehghan B, Farahani MA. (2019). Health-related quality of life of breast cancer patients in the Eastern Mediterranean region: A systematic review and meta-analysis. *Breast Cancer Res Treat*, **174**(3): 585–596.

4. Siegel RL, Miller KD, Jemal A. (2018). Cancer statistics, 2018. *CA Cancer J Clin*, **68**(1): 7–30.

5. Althuis MD, Dozier JM, Anderson WF, Devesa SS, Brinton LA. (2005). Global trends in breast cancer incidence and mortality 1973–1997. *Int J Epidemiol*, **34**(2): 405–412.

6. Bray F, McCarron P, Parkin DM. (2004). The changing global patterns of female breast cancer incidence and mortality. *Breast Cancer Res*, **6**(6): 229–239.

7. Vuong D, Simpson PT, Green B, Cummings MC, Lakhani SR. (2014). Molecular classification of breast cancer. *Virchows Arch*, **465**(1): 1–14.

8. de Ruijter TC, Veeck J, de Hoon JP, van Engeland M, Tjan-Heijnen VC. (2011). Characteristics of triple-negative breast cancer. *J Cancer Res Clin Oncol,* **137**(2): 183–192.

9. Perou CM, Sorlie T, Eisen MB, van de Rijn M, Jeffrey SS, Rees CA, Pollack JR, Ross DT, Johnsen H, Akslen LA *et al.* (2000). Molecular portraits of human breast tumours. *Nature,* **406**(6797): 747–752.

10. Sorlie T, Perou CM, Tibshirani R, Aas T, Geisler S, Johnsen H, Hastie T, Eisen MB, van de Rijn M, Jeffrey SS *et al.* (2001). Gene expression patterns of breast carcinomas distinguish tumor subclasses with clinical implications. *Proc Natl Acad Sci USA,* **98**(19): 10869–10874.

11. Prat A, Parker JS, Karginova O, Fan C, Livasy C, Herschkowitz JI, He X, Perou CM. (2010). Phenotypic and molecular characterization of the claudin-low intrinsic subtype of breast cancer. *Breast Cancer Res,* **12**(5): R68.

12. Gregorio AC, Lacerda M, Figueiredo P, Simoes S, Dias S, Moreira JN. (2018). Therapeutic implications of the molecular and immune landscape of triple-negative breast cancer. *Pathol Oncol Res,* **24**(4): 701–716.

13. Zhao S, Ma D, Xiao Y, Li XM, Ma JL, Zhang H, Xu XL, Lv H, Jiang WH, Yang WT *et al.* (2020). Molecular subtyping of triple-negative breast cancers by immunohistochemistry: Molecular basis and clinical relevance. *Oncologist.*

14. Millikan RC, Newman B, Tse CK, Moorman PG, Conway K, Dressler LG, Smith LV, Labbok MH, Geradts J, Bensen JT *et al.* (2008). Epidemiology of basal-like breast cancer. *Breast Cancer Res Treat,* **109**(1): 123–139.

15. Carey L, Winer E, Viale G, Cameron D, Gianni L. (2010). Triple-negative breast cancer: Disease entity or title of convenience? *Nat Rev Clin Oncol,* 7(12): 683–692.

16. Hammond ME, Hayes DF, Dowsett M, Allred DC, Hagerty KL, Badve S, Fitzgibbons PL, Francis G, Goldstein NS, Hayes M *et al.* (2010). American society of clinical oncology/college of american pathologists guideline recommendations for immunohistochemical testing of estrogen and progesterone receptors in breast cancer. *J Clin Oncol,* **28**(16): 2784–2795.

17. Nanda R. (2011). "Targeting" triple-negative breast cancer: The lessons learned from BRCA1-associated breast cancers. *Semin Oncol,* **38**(2): 254–262.

18. Bertucci F, Finetti P, Cervera N, Esterni B, Hermitte F, Viens P, Birnbaum D. (2008). How basal are triple-negative breast cancers? *Int J Cancer,* **123**(1): 236–240.

19. Rakha EA, Elsheikh SE, Aleskandarany MA, Habashi HO, Green AR, Powe DG, El-Sayed ME, Benhasouna A, Brunet JS, Akslen LA *et al.* (2009). Triple-negative breast cancer: Distinguishing between basal and nonbasal subtypes. *Clin Cancer Res,* 15(7): 2302–2310.

20. Anders CK, Carey LA. (2009). Biology, metastatic patterns, and treatment of patients with triple-negative breast cancer. *Clin Breast Cancer,* 9 Suppl 2: S73–81.

21. Wooster R, Weber BL. (2003). Breast and ovarian cancer. *N Engl J Med,* 348(23): 2339–2347.

22. Venkitaraman AR. (2002). Cancer susceptibility and the functions of BRCA1 and BRCA2. *Cell,* 108(2): 171–182.

23. Foulkes WD, Stefansson IM, Chappuis PO, Begin LR, Goffin JR, Wong N, Trudel M, Akslen LA. (2003). Germline BRCA1 mutations and a basal epithelial phenotype in breast cancer. *J Natl Cancer Inst,* 95(19): 1482–1485.

24. Wilson CA, Ramos L, Villasenor MR, Anders KH, Press MF, Clarke K, Karlan B, Chen JJ, Scully R, Livingston D *et al.* (1999). Localization of human BRCA1 and its loss in high-grade, non-inherited breast carcinomas. *Nat Genet,* 21(2): 236–240.

25. The Cancer Genome Atlas N, Koboldt DC, Fulton RS, McLellan MD, Schmidt H, Kalicki-Veizer J, McMichael JF, Fulton LL, Dooling DJ, Ding L *et al.* (2012). Comprehensive molecular portraits of human breast tumours. *Nature,* 490: 61.

26. Thompson D, Easton DF. (2002). Breast Cancer Linkage C: Cancer incidence in BRCA1 mutation carriers. *J Natl Cancer Inst,* 94(18): 1358–1365.

27. Plasilova ML, Hayse B, Killelea BK, Horowitz NR, Chagpar AB, Lannin DR. (2016). Features of triple-negative breast cancer: Analysis of 38,813 cases from the national cancer database. *Medicine (Baltimore),* 95(35): e4614.

28. Bauer KR, Brown M, Cress RD, Parise CA, Caggiano V. (2007). Descriptive analysis of estrogen receptor (ER)-negative, progesterone receptor (PR)-negative, and HER2-negative invasive breast cancer, the so-called triple-negative phenotype: A population-based study from the California cancer Registry. *Cancer,* 109(9): 1721–1728.

29. Bowen RL, Duffy SW, Ryan DA, Hart IR, Jones JL. (2008). Early onset of breast cancer in a group of British black women. *Br J Cancer,* 98(2): 277–281.

30. Huo D, Ikpatt F, Khramtsov A, Dangou JM, Nanda R, Dignam J, Zhang B, Grushko T, Zhang C, Oluwasola O *et al.* (2009). Population

differences in breast cancer: Survey in indigenous African women reveals over-representation of triple-negative breast cancer. *J Clin Oncol,* **27**(27): 4515–4521.

31. Carey LA, Perou CM, Livasy CA, Dressler LG, Cowan D, Conway K, Karaca G, Troester MA, Tse CK, Edmiston S *et al.* (2006). Race, breast cancer subtypes, and survival in the Carolina Breast Cancer Study. *JAMA,* **295**(21): 2492–2502.

32. Gucalp A, Traina TA, Eisner JR, Parker JS, Selitsky SR, Park BH, Elias AD, Baskin-Bey ES, Cardoso F. (2019). Male breast cancer: A disease distinct from female breast cancer. *Breast Cancer Res Treat,* **173**(1): 37–48.

33. Dent R, Trudeau M, Pritchard KI, Hanna WM, Kahn HK, Sawka CA, Lickley LA, Rawlinson E, Sun P, Narod SA. (2007). Triple-negative breast cancer: Clinical features and patterns of recurrence. *Clin Cancer Res,* **13**(15 Pt 1): 4429–4434.

34. Foulkes WD, Smith IE, Reis-Filho JS. (2010). Triple-negative breast cancer. *N Engl J Med,* **363**(20): 1938–1948.

35. Bianchini G, Balko JM, Mayer IA, Sanders ME, Gianni L. (2016). Triple-negative breast cancer: Challenges and opportunities of a heterogeneous disease. *Nat Rev Clin Oncol,* **13**(11): 674–690.

Chapter TWO

Heterogeneity and Subtyping of Triple-Negative Breast Cancer

Yingying Wu[1], Qiuting Zhang[2], *and* Ceshi Chen[1,*]

Contents

---

*Corresponding author: Ceshi Chen, E-mail: chenc@mail.kiz.ac.cn

[1] Key Laboratory of Animal Models and Human Disease Mechanisms of the Chinese Academy of Sciences and Yunnan Province, Kunming Institute of Zoology, Chinese Academy of Sciences, Kunming, China.

[2] Key Laboratory of Translational Cancer Stem Cell Research, Hunan Normal University, Changsha, Hunan, China.

2.1 Introduction

Since the response rates for the targeted therapies vary enormously from patients to patients, it is postulated that triple-negative breast cancer (TNBC) is a heterogenous subtype of breast cancer. Indeed, TNBC has been increasingly recognized as a heterogeneous disease that exhibits substantial differences in terms of genomic, transcriptomic, and perhaps proteomic profiles. The extreme heterogeneity of TNBC has led to difficulties in finding suitable molecular targets in preclinical studies. Most targeted agents tested so far, except for PARP inhibitors and immune checkpoint blockers (discussed in Chapters 6 and 7), have demonstrated low overall activity in unselected TNBC. These limited benefits from targeted therapies further highlight the molecular heterogeneity of TNBC and suggest the importance of subtype-specific treatment for TNBC patients. Therefore, TNBC subtyping based on the biologically and clinically relevant characteristics may contribute to the identification of therapeutic targets, optimization of clinical trial designs, and patient risk stratification [1].

This concept of molecular heterogeneity is beginning to be accepted by researchers and clinicians and refined by TNBC's molecular characteristics and clinical response to currently available therapies. The treatment paradigm of "one size fits all" approach for the management of TNBC is changing based on histopathological and molecular subtyping. The inter-tumoral heterogeneity of TNBC is an important factor accounting for the poor results of many of these efforts in unselected TNBC patients. This chapter describes about our current understanding of TNBC as a heterogenous disease entity with clinically and pathologically significant subtyping.

2.2 Clinical Heterogeneity of TNBC

TNBC has been shown to have a high prevalence in the women of African and Hispanic descent at an early age (under 40 years) of presentation [2], and has been characterized by aggressive progression and poor survival compared with other breast cancer subtypes. Although most TNBCs are high-grade tumors with a relatively poor prognosis, a subset of low-grade TNBCs display a favorable outcome [3–5].

TNBC displays a specific pattern of relapse. These tumors have a predilection for visceral, lung and brain metastasis, whereas luminal breast cancers favor relapses in bone and skin [6–8]. At the clinical level, distant recurrences peak early at 3 years following diagnosis and a majority of deaths occur in the first 5 years after initial diagnosis [9]. Interestingly, there are similar survival rates between patients with TNBC who have not recurred during this time and patients with ER-positive breast cancers [10, 11].

Despite rather malignant biological properties of some TNBC tumors, patients with TNBC are more sensitive to initial anthracyclines (such as doxorubicin) and taxanes (such as paclitaxel) compared with other breast cancer subtypes, and the clinical response rates and pathologic complete response (pCR) rates can be up to 85% and 30–40%, respectively [12]. When the patients were treated with neoadjuvant chemotherapy before surgery, the disease-free survival (DFS) and overall survival (OS) of patients with pCR have significant improvements compared to patients with residual invasive disease [13].

2.3 Histopathological Subtyping of TNBC

2.3.1 *Histopathological Subtypes of TNBC*

TNBC encompasses a variety of histopathological types. Invasive carcinoma of no special type, referred to as NST, is by far the vast majority of TNBC, which accounts for >80% of all TNBC cases. Other histologic types, which are referred to as special histologic types, account for approximately 10% of all TNBCs [14, 15]. Invasive carcinoma of no special type represents a heterogeneous group of tumors that fail to exhibit sufficient histopathological features to be classified as any special type. The morphological features of this subtype vary vastly among different cases. The most common special histopathological types of TNBC include metaplastic carcinoma, medullary carcinoma, invasive lobular carcinoma, apocrine carcinoma and adenoid cystic carcinoma. Several studies have reported the prognostic implications of special histological types in TNBC. Compared with invasive carcinoma of no special type, metaplastic and invasive lobular carcinomas are associated with a poorer prognosis, while medullary, apocrine and adenoid cystic carcinomas dictate a better prognosis [1].

Although classification of TNBC based on histopathological features has been well described in the literature, very little treatment recommendations are made in current guidelines according to histopathological typing. The prognostic implications of certain histologic types may be valuable in making clinical decisions regarding patient follow-up and therapeutic approaches, for example the dose intensity and the duration of adjuvant chemotherapy.

2.3.2 *Major Histopathological Subtypes of TNBC*

Histopathologically, TNBC is mainly divided into three categories: (1) high-grade invasive ductal carcinomas of no special type; (2) high-grade special histologic types of breast cancer, including carcinomas with apocrine features, carcinomas with medullary features, and MBCs; (3) low-grade TNBC which can be further classified in at least two subgroups including salivary gland-like tumors of breast and low-grade TN breast neoplasia family. The salivary gland-like tumors of breast comprise adenoid cystic carcinoma and secretory carcinoma. The low-grade TN breast neoplasia family comprises microglandular adenosis, atypical microglandular adenosis and acinic cell carcinoma [16].

2.3.2.1 *High-grade invasive ductal carcinomas of no special type*

As mentioned above, high-grade invasive ductal carcinomas of no special type is the most common type, accounting for the vast majority (>80%) of all TNBC cases. The morphological characteristics are high histopathological tumor grade, marked cellular pleomorphism, lack of tubule formation, brisk lymphocyte infiltration, scant stromal content, pushing edge of invasion, central geographic or comedotype necrosis, and central acellularity [17]. The majority of this group of TNBC are basal-like (BL) subtype characterized by the expression of myoepithelial/basal markers and molecular changes including TP53 gene mutations, BRCA1 inactivation, and chromosomal alterations [18]. The 5-year OS rate for this subtype of TNBC was 62% [19].

2.3.2.2 *Metaplastic breast carcinomas*

Metaplastic breast carcinoma is the special histologic type accounting for 4% of TNBCs [19]. The morphological characteristics are displaying differentiation towards squamous epithelium with mesenchymal components and cells having spindle, chondroid, osseous, or rhabdoid morphologies [20]. The hallmark features of metaplastic breast carcinoma are worse outcome than conventional TNBC [21], showing significant inter- and intra-tumor heterogeneity [22]. The enrichment of genetic alterations involves Wnt and PI3K pathways [23, 24], particularly PIK3CA mutations.

2.3.2.3 *Carcinomas with medullary features*

It is the special histologic type accounting for 2.3% of TNBCs [19]. The morphological characteristics are well-circumscribed borders, a syncytial growth pattern, and brisk lymphocytic infiltrate. This subtype of TNBC is characterized by excellent outcome, although with its worrisome cytological features and high mitotic activity [20]. The 5-year OS rate was 100% for these patients [19].

2.3.2.4 *Carcinomas with apocrine features*

It is another special histologic type accounting for 1.6–3.7% of TNBC [19, 25]. The morphological characteristics are abundant eosinophilic cytoplasm and prominent nucleoli. These tumors are most likely of LAR subtype characterized by expressing androgen receptor (AR) and displaying a molecular apocrine or LAR gene expression profile [26]. The outcome of this subtype of TNBC is uncertain, because contradictory data have been published regarding the prognostic impact of AR expression in TNBC [27, 28].

2.3.2.5 *Low-grade TNBC*

Low-grade TNBCs can be further classified in at least two subgroups including salivary gland-like tumors of the breast and low-grade TN breast neoplasia family, encompass microglandular adenosis, atypical microglandular adenosis and acinic cell carcinoma. These two subgroups have been shown to have low-grade morphology and indolent clinical behaviors [16]. Even among metaplastic breast carcinomas characterized by high-grade lesions and worse outcome than conventional TNBC [21], low-grade variants exist, such as the low-grade spindle and adenosquamous carcinomas, which display relatively better clinical outcome [29].

2.3.3 *The Underlying Pattern of Progression of TNBC*

Conventional high-grade TNBCs usually derive from normal breast epithelium when acquiring TP53 and PIK3CA mutations. However, there are other potential evolutionary paths of TNBCs. Two subtypes of low-grade TNBCs have been known as the basis or precursors of TNBCs. Furthermore, both low-grade subgroups can progress to high-grade TNBCs. Salivary gland-like tumors of the breast are evolved into high-grade TNBCs via the acquisition of additional genetic events and/or clonal selection. Notably, high-grade TNBCs arising in salivary gland-like tumors are unlike with conventional TNBCs at the genetic level, similar to their respective low-grade counterparts. There's a higher chance of transforming into high-grade TNBCs for the other subgroup, the so-called low-grade TN breast neoplasia family, which are similar to conventional TNBCs in phenotype and genetic alterations [16] (**Figure 2-1**).

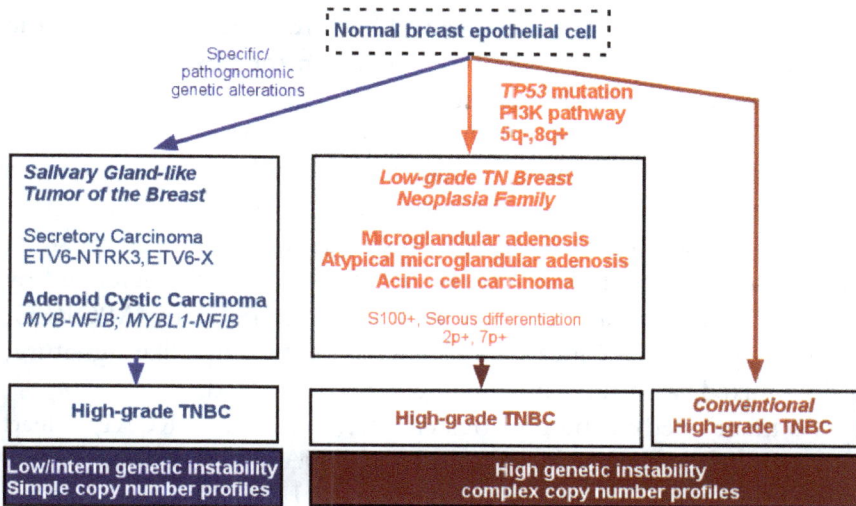

Figure 2-1. Hypothetical model of potential paths of TNBC progression. Conventional high-grade TNBCs are usually derived from normal breast epithelial cells when acquiring TP53 and PIK3CA mutations. There are two subtypes of low-grade TNBCs, which both can progress to high-grade TNBCs. The high-grade TNBCs derived from low-grade triple-negative breast neoplasia family accord with conventional TNBCs in phenotype and genetic alterations, whereas high-grade TNBCs arising in salivary gland-like tumors are different from conventional TNBCs at the genetic level.

2.4 Molecular Subtyping of TNBC

2.4.1 *Subtyping Based on Genomic Alterations*

Cancer genomes harbor a large amount of somatic mutations, but only a few of them play a role in driving carcinogenesis by conferring selective advantage to tumor cell growth. Genomic profiling studies of TNBC, including whole-exome and whole-genome analyses, have identified several recurrent alterations in the so-called cancer driver genes. Somatic mutations and copy number alterations (CNAs) account for most genomic alterations in TNBC. TP53 is the most frequently mutated gene, followed by PIK3CA, PTEN, KMT2C, and RB1. MYC amplification is the most frequent CNA event in TNBC. Other genes frequently affected by somatic CNAs include EGFR,

PTEN, CCND1, RB1, and CCNE1 [1]. More information about the genetic changes in TNBC will be discussed in the next chapter.

Different mutational processes often generate different combinations of mutation types, which is termed "signature". Nik-Zainal *et al.* analyzed whole genome sequences of 560 breast cancer samples and identified twelve base substitution and six rearrangement signatures [30]. Based on these established signatures, Jiang *et al.* classified TNBC into four mutation subtypes [31]: 1) HRD, characterized by HRD-related signature; 2) APOBEC, characterized by APOBEC-related signature; 3) Clock-like, characterized by clock-like signature; and 4) mixed, with no dominant signature identified.

As mentioned in the previous chapter, defects in BRCA1/2 lead to a type of deficiency in the repair of DNA double-strand breaks called homologue recombination deficiency (HRD). Previous studies reported that 10–20% TNBCs have germline mutations in BRCA1/2 genes. These tumors typically exhibit HRD, but some sporadic (BRCA1/2 germline wild-type) TNBCs can also display functional BRCA1/2 deficiency and harbor DNA repair defects. Thus, researchers sought to uncover additional biomarkers indicative of HRD beyond germline BRCA1/2 mutations. Three quantitative metrics, i.e., loss of heterozygosity (LOH) [32], telomeric allelic imbalance (TAI) [33], and large-scale state transitions (LST) [34], have been developed to measure the genomic instability that is the consequence of HRD. A combined HRD score is defined as the arithmetic mean of these three scores. In the neoadjuvant setting, either HRD in TNBC or a high HRD score in the BRCA1/2 wild-type subgroup predicted better response towards platinum [35, 36]. However, in the metastatic setting, different groups reported inconsistent results.

By analyzing the whole genome sequencing data of 560 breast cancer samples, Scientists from Wellcome Trust Sanger Institute, Hinxton, UK developed a mutational signature-based predictor of BRCA1/2 deficiency called HRDetect [37]. Later, the same group applied the HRDetect algorithm to 254 TNBC samples and classified them into HRDetect-high, -intermediate, and -low subgroups. The HRDetect-high subgroup had a high degree of sensitivity to standard adjuvant chemotherapy and was associated with a better prognosis

[38]. With the advancement in sequencing technology and reduced cost, the HRDetect model may be used in the future to inform trial stratification and improve clinical outcomes of TNBC.

2.4.2 *Subtyping Based on Gene Expression Profiling*

Advances in gene expression analyses and clinical sequencing allow to classify TNBCs into further molecular subtypes. Currently, there are several transcriptome-based classification methods for TNBC, but the major subtyping methods include Vanderbilt subtyping, Baylor subtyping, and Fudan subtyping [39, 40].

2.4.2.1 *Vanderbilt subtyping*

In 2011, Lehmann *et al.* first initiated molecular subtyping of TNBC into six subtypes based on the PAM50 gene expression profiling: basal-like 1 (BL1), BL2, immunomodulatory (IM) subtype, mesenchymal (M) subtype, mesenchymal stem-like (MSL) subtype, and luminal androgen receptor (LAR) subtype [41]. More importantly, they identified different cell lines representing corresponding TNBC subtypes by analyzing distinct gene expression profiles, and predicted that different activated signaling pathways could be pharmacologically targeted in cell lines [39]. BL-TNBC is characterized by DNA-repair deficiency, and the cisplatin treatment is effective. The M and MSL subtypes display higher expression of genes involved in EMT and activation of receptor tyrosine kinase pathways, and inhibitors of PI3K/mTOR and ABL/SRC are effective in representative cell lines. The LAR subtype is characterized by AR signaling and shows luminal gene expression pattern. AR antagonists are effective in LAR cell lines [42]. In 2016, these subtypes have been refined to four distinct subtypes: BL1 and BL2, M, and LAR (**Figure 2-2**), because the transcripts in the previously described IM and MSL subtypes were found to be derived from infiltrating lymphocytes and tumor-associated stromal cells, respectively [43].

In order to determine the potential clinical utility of classifying tumors by TNBC subtype, Masuda *et al.* [44] performed a

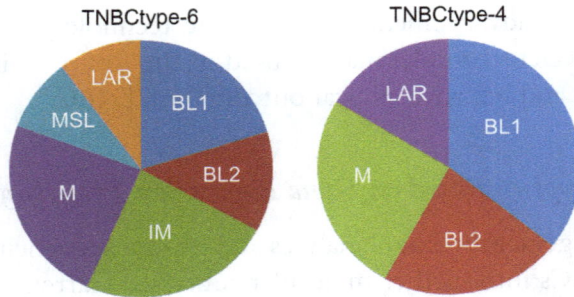

Figure 2-2. Molecular subtypes of TNBC stratified by PAM50 gene expression profiling. The proportions of the original TNBCtype-6 (left) and the refined TNBCtype-4 (right) subtypes of TNBC are depicted.

retrospective analysis on 130 TNBC patients treated with neoadjuvant chemotherapy (anthracycline and taxane-based). The results showed that while the overall pCR rate was 28%, subtype-specific responses differed substantially. The BL1 subtype achieved the highest pCR rate (52%), whereas the BL2, LAR, and MSL subtypes had the poorer response (0%, 10%, and 23%, respectively). Nevertheless, a larger number of TNBC cases are needed to ascertain the association between the molecular subtypes and clinical outcomes of TNBC.

2.4.2.2 Baylor subtyping

Using similar PAM50 gene expression profiling, Burstein *et al.* from Baylor College of Medicine classified TNBCs into four subtypes: LAR, M, basal-like immune-suppressed (BLIS), and basal-like immune-activated (BLIA) [45]. Further studies demonstrated a high correlation between the Vanderbilt BL1/BL2 and the Baylor BLIA/BLIS subtypes. They also identified putative subtype-specific targets: AR, ER (although ER negative by immunohistochemistry), prolactin, and cell-surface MUC (MUC-1) for the LAR subtype; platelet-derived growth factor (PDGF) receptor A, insulin-like growth factor 1 (IGF1), and c-Kit for the M subtype; Sry-related HMG box (SOX) transcription factors as well as V-set domain-containing T-cell activation inhibitor 1 (VTCN1) for the BLIS subtype; signal transducer and activator of transcription (STAT), cytotoxic T lymphocyte-associated

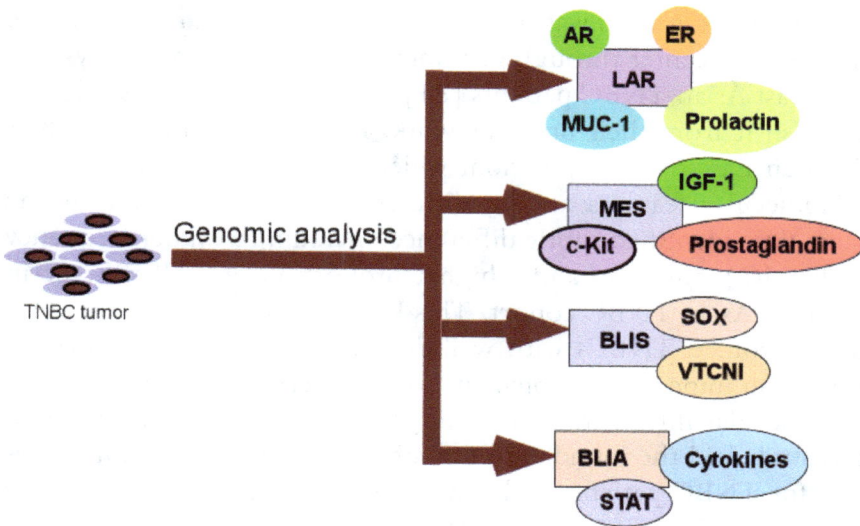

Figure 2-3. The genomic landscape of TNBC according to the Baylor subtypes. Four distinct subtypes of TNBC with specific molecular targets are shown.

antigen 4 (CTLA4) and cytokines for the BLIA subtype. These studies imply a promising future for personalized therapy in TNBC based on molecular subtyping [45] (**Figure 2-3**).

2.4.2.3 *Fudan subtyping*

In 2016, using a novel classification system integrating transcriptome profiling of mRNA and lncRNA, Liu *et al.* from Fudan University Shanghai Cancer Center (FUSCC) classified TNBC into four distinct clusters: IM, LAR, mesenchymal-like (MES) and basal-like immune suppressed (BLIS) subtype. The IM subtype was characterized by high expression of immune cell signaling and cytokine signaling genes, which is similar to the Baylor BLIA subtype. The LAR subtype displayed AR signaling activation. The MES subtype was enriched in breast cancer stem cell pathways. The BLIS subtype was featured by downregulation of immune response genes, activation of cell cycle and DNA repair. Patients with the BLIS subtype had the worst recurrence-free survival than other subtypes [46].

This classification was further confirmed by Jiang *et al.* from the same cancer center through genomic and transcriptomic analyses of 465 East Asian TNBC patients [31]. Two aspects of this study distinguish it from their previous work. First, they analyzed TNBCs from an East Asian population, which demonstrated the similarity in molecular features among different ethnic groups and at the same time identified subtle difference, including a higher frequency of PIK3CA mutation and a higher proportion of LAR subtype in the East Asian TNBC cohort. This large collection of comprehensively profiled TNBCs with well-documented clinical information will be an important supplement to the international compendium of molecular information regarding human breast cancer. Second, this study laid the foundation for subtype-specific treatment strategies for TNBC patients and a subsequent clinical trial is currently underway (NCT3805399) pending publication of preliminary results [1].

Different biomarkers and treatment strategies were proposed for these four subtypes of TNBC. More PIK3CA mutations, activated HER2/ErbB2 and cell cycle signaling were identified in LAR subtype, suggesting that CDK 4/6 inhibitors may be effective in this subtype. Immune checkpoint blockade is the promising therapeutic approach for the IM subtype. Patients with high HRD score may benefit substantially from DNA-damaging therapies, such as platinum-based chemotherapy or PARP inhibitors, a phenomenon known as "synthetic lethality" (discussed in Chapter 5 and Chapter 6). Finally, JAK/STAT3 signaling pathway is activated in the MES subtype, which gives the opportunity for these patients to potentially benefit from JAK/STAT inhibitors, such as ruxolitinib.

Additionally, TNBC can be classified according to the tumor microenvironment characteristics, such as the abundance of tumor-infiltrating cells (TILs). Accordingly, TNBCs can be either "immune-hot" or "immune-cold", depending on the level of TILs. This will be discussed in Chapter 8. The intrinsic connection of the three major transcriptome-based TNBC subtyping methods is shown in **Figure 2-4**.

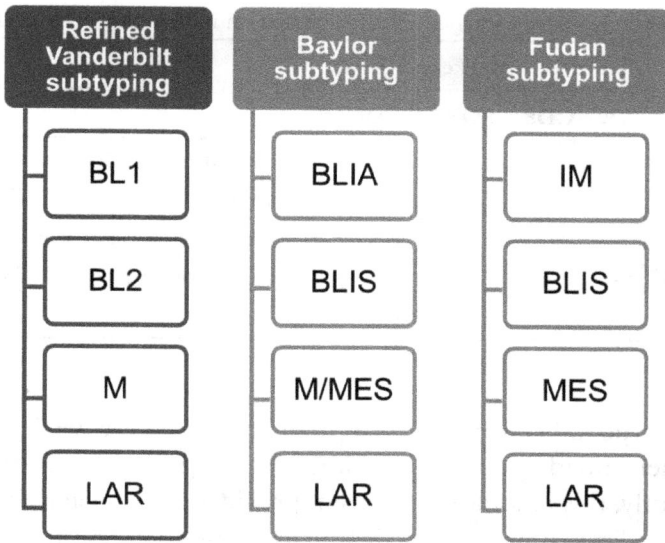

Figure 2-4. Correlations between Vanderbilt, Baylor, and Fudan subtyping. Intrinsic connection of the three major transcriptome-based TNBC subtyping methods.

2.4.3 *Subtyping Based on Immunohistochemistry*

To establish a clinically feasible immunohistochemistry (IHC)-based classification of TNBC, Zhao *et al.* first analyzed the RNA sequencing data on TNBCs from FUSCC and The Cancer Genome Atlas (TCGA) dataset and determined markers that can be used to identify specific molecular subtypes of TNBC [47]. Five subtypes were classified based on the immunohistochemical staining results: IHC-LAR (AR+), IHC-IM (AR-, CD8+), IHC-BLIS (AR-, CD8-, FOXC1+), IHC-MES (AR-, CD8-, FOXC1-, DCLK1+), and IHC-UC for unclassifiable samples (AR-, CD8-, FOXC1-, DCLK1-) (**Table 2-1**). The IHC-LAR subtype showed relative activation of the HER2 pathway. The IHC-IM subtype tended to exhibit an "immune-inflamed" phenotype (refer to Chapter 8 for description of "immune-inflamed" tumors) characterized by the infiltration of CD8+ T cells into tumor parenchyma. The IHC-BLIS subtype showed high expression of a VEGF signature. The IHC-MES subtype displayed activation of the JAK/STAT3 signaling pathway.

Table 2-1.   Five subtypes of TNBC based on immunohistochemistry.

Subtype	Markers				Characteristics
	AR	CD8	FOXC1	DCLK1	
IHC-LAR	+				Activation of HER2 pathway
IHC-IM	−	+			Infiltration of CD8+ T cells
IHC-BLIS	−	−	+		High expression of VEGF signature
IHC-MES	−	−	−	+	Activation of JAK/STAT3 pathway
IHC-UC	−	−	−	−	

This type of classification approach agrees very well with the above-mentioned gene expression profiling-based classification. Importantly, this classification system provides additional information beyond traditional prognostic factors in relapse prediction and allows for subgroup-specific targeted therapies of TNBC patients in large-scale clinical trials [47].

2.4.4 *Correlation Between Molecular and Histopathological Subtypes and Clinical Implications*

Some molecular subtypes have significant correlation with the histo-pathological subtypes as sharing similar gene expression signatures. For example, the majority of high-grade invasive carcinomas of no special type are BL subtypes (BL1 and BL2) characterized by the expression of myoepithelial/basal markers and molecular changes including TP53 gene mutations and BRCA1 inactivation [18]. The gene expression of M subtype overlaps largely with the metaplastic cancers [48, 49]. In addition, the LAR subtype corresponds with the apocrine and rare cancers (**Figure 2-5**).

Despite the differences in the classification methods and nomenclature in the literature, the classification results showed obvious agreement between each other. International effort to sharing and merging large-scale data and cross-comparison of different classification approaches may lead to a final consensus. Transcriptome-based profiling analysis provides great insight into the molecular heterogeneity of TNBC and enables robust and unbiased classification to

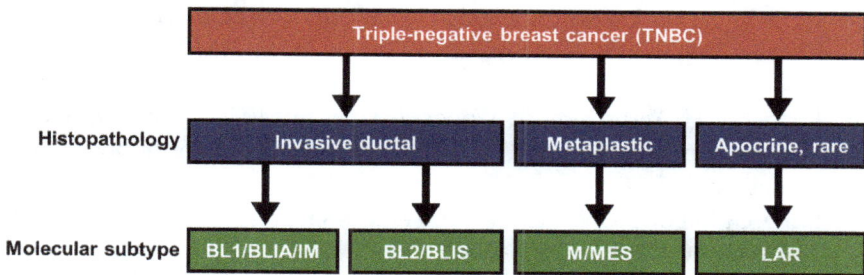

Figure 2-5. Correlations between histopathology and molecular subtyping in TNBC. See text for details. BL1, basal-like 1; BL2, basal-like 2; BLIA, basal-like immune-activated; BLIS, basal-like immune-suppressed; IM, immunomodulatory; M, mesenchymal; LAR, luminal androgen receptor.

direct clinical decision-making efforts. The IHC-based approach can be a feasible and easy-to-perform classification system hopefully to guide treatment decisions for patients with TNBC in the near future.

References

1. Zhao S, Zuo WJ, Shao ZM, Jiang YZ. (2020). Molecular subtypes and precision treatment of triple-negative breast cancer. *Ann Transl Med*, 8(7): 499–513.
2. Amirikia KC, Mills P, Bush J, Newman LA. (2011). Higher population-based incidence rates of triple-negative breast cancer among young African-American women implications for breast cancer screening recommendations. *Cancer*, 117(12): 2747–2754.
3. Guerini-Rocco E, Hodi Z, Piscuoglio S, Ng CK, Rakha EA, Schultheis AM, Marchiò C, Cruz Paula AD, De Filippo MR, Martelotto LG. (2015). The repertoire of somatic genetic alterations of acinic cell carcinomas of the breast: An exploratory, hypothesis-generating study. *J Pathol*, 237(2): 166–178.
4. Del CM, Chibon F, Arnould L, Croce S, Ribeiro A, Perot G, Hostein I, Geha S, Bozon C, Garnier A. (2015). Secretory breast carcinoma: A histopathologic and genomic spectrum characterized by a joint specific ETV6-NTRK3 gene fusion. *Am J Surg Pathol*, 39(11): 1458–1467.
5. Fusco N, Geyer FC, Filippo MRD, Martelotto LG, Ng CKY, Piscuoglio S, Guerinirocco E, Schultheis AM, Fuhrmann L, Wang L. (2016).

Genetic events in the progression of adenoid cystic carcinoma of the breast to high-grade triple-negative breast cancer. *Mod Pathol*, **29**(11): 1292–1305.

6. Kennecke H, Yerushalmi R, Woods R, Cheang MCU, Voduc D, Speers CH, Nielsen TO, Gelmon K. (2010). Metastatic behavior of breast cancer subtypes. *J Clin Oncol*, **28**(20): 3271–3277.

7. Lin NU, Bellon JR, Winer EP. (2004). CNS metastases in breast cancer. *J Clin Oncol*, **22**(17): 3608–3617.

8. Heitz F, Harter P, Beutel B, Lueck HJ, Traut A, Bois AD. (2008). Cerebral metastases (CM) in breast cancer (BC) with focus on "triple-negative" (TN) tumors. *Geburtshilfe Frauenheilkd*, **68**(S 01): 431–436.

9. Dent R, Trudeau M, Pritchard K, Hanna W, Kahn H, Sawka C, Lickley L, Rawlinson E, Sun P, Narod S. (2007). Triple-negative breast cancer: Clinical features and patterns of recurrence. *Clin Cancer Res*, **13**(15 Pt 1): 4429–4434.

10. Liedtke C, Mazouni C, Hess KR, André F, Tordai A, Mejia JA, Symmans WF, Gonzalez-Angulo AM, Hennessy B, Green M *et al.* (2008). Response to neoadjuvant therapy and long-term survival in patients with triple-negative breast cancer. *J Clin Oncol*, **26**(8): 1275–1281.

11. Dent R, Trudeau M, Pritchard KI, Hanna WM, Kahn HK, Sawka CA, Lickley LA, Rawlinson E, Sun P, Narod SA. (2007). Triple-negative breast cancer: Clinical features and patterns of recurrence. *Clin Cancer Res*, **13**(15 Pt 1): 4429–4434.

12. Von Minckwitz G, Untch M, Blohmer JU, Costa SD, Eidtmann H, Fasching PA, Gerber B, Eiermann W, Hilfrich J, Huober J *et al.* (2012). Definition and impact of pathologic complete response on prognosis after neoadjuvant chemotherapy in various intrinsic breast cancer subtypes. *J Clin Oncol*, **30**(15): 1796–1804.

13. Hage JHVD, Velde CCVD, Mieog SJ. (2007). Preoperative Chemotherapy for Women with Operable Breast Cancer: John Wiley & Sons, Ltd.

14. Foulkes WD, Smith IE, Reis-Filho JS. (2010). Triple-negative breast cancer. *N Engl J Med*, **363**(20): 1938–1948.

15. Turner NC, Reis-Filho JS. (2013). Tackling the diversity of triple-negative breast cancer. *Clin Cancer Res*, **19**(23): 6380–6388.

16. Pareja F, Geyer FC, Marchiò C, Burke KA, Weigelt B, Reisfilho JS. (2016). Triple-negative breast cancer: The importance of molecular and histologic subtyping, and recognition of low-grade variants. *Npj Breast Cancer*, **2**: 16036–16047.

17. Rakha EA, Reisfilho JS, Ellis IO. (2008). Basal-like breast cancer: A critical review. *J Clin Oncol*, **26**(15): 2568–2581.
18. Sasaki Y, Tsuda H. (2009). Clinicopathological characteristics of triple-negative breast cancers. *Breast Cancer*, **16**(4): 254–259.
19. Cao L, Niu Y. (2020). Triple negative breast cancer: Special histological types and emerging therapeutic methods. *Cancer Biol Med*, **17**(2): 293–306.
20. Huober J, Gelber S, Goldhirsch A, Coates AS, Viale G, Öhlschlegel C, Price KN, Gelber RD, Regan MM, Thürimann B. (2012). Prognosis of medullary breast cancer: Analysis of 13 International Breast Cancer Study Group (IBCSG) trials. *Ann Oncol*, **23**(11): 2843–2851.
21. Jung SY, Kim HY, Nam BH, Min SY, Lee SJ, Park C, Kwon Y, Kim EA, Ko KL, Shin KH. (2010). Worse prognosis of metaplastic breast cancer patients than other patients with triple-negative breast cancer. *Breast Cancer Res Treat*, **120**(3): 627–637.
22. Geyer FC, Kushner YB, Lambros MB, Natrajan R, Mackay A, Tamber N, Fenwick K, Purnell D, Ashworth A, Walker RA. (2010). Microglandular adenosis or microglandular adenoma? A molecular genetic analysis of a case associated with atypia and invasive carcinoma. *Histopathology*, **55**(6): 732–743.
23. Hennessy BT, Gonzalez-Angulo AM, Stemke-Hale K, Gilcrease MZ, Krishnamurthy S, Lee JS, Fridlyand J, Sahin A, Agarwal R, Joy C *et al.* (2009). Characterization of a naturally occurring breast cancer subset enriched in epithelial-to-mesenchymal transition and stem cell characteristics. *Cancer Res*, **69**(10): 4116–4124.
24. Hayes MJ, Thomas D, Emmons A, Giordano TJ, Kleer CG, Hayes MJ, Thomas D, Emmons A, Giordano TJ, Kleer CG. (2008). Genetic changes of wnt pathway genes are common events in metaplastic carcinomas of the breast. *Clin Cancer Res*, **14**(13): 4038–4044.
25. Montagna E, Maisonneuve P, Rotmensz N, Cancello G, Iorfida M, Balduzzi A, Galimberti V, Veronesi P, Luini A, Pruneri G. (2013). Heterogeneity of triple-negative breast cancer: Histologic subtyping to inform the outcome. *Clin Breast Cancer*, **13**(1): 31–39.
26. Farmer P, Bonnefoi H, Becette V, Tubiana-Hulin M, Fumoleau P, Larsimont D, Macgrogan G, Bergh J, Cameron D, Goldstein D *et al.* (2005). Identification of molecular apocrine breast tumours by microarray analysis. *Oncogene*, **24**(29): 4660–4671.
27. Choi JE, Su HK, Lee SJ, Bae YK. (2015). Androgen receptor expression predicts decreased survival in early stage triple-negative breast cancer. *Ann Surg Oncol*, **22**(1): 82–89.

28. Vera-Badillo FE, Templeton AJ, de Gouveia P, Diaz-Padilla I, Bedard PL, Al-Mubarak M, Seruga B, Tannock IF, Ocana A, Amir E. (2014). Androgen receptor expression and outcomes in early breast cancer: A systematic review and meta-analysis. *J Natl Cancer Inst*, **106**(1): djt319.

29. Lakhani, Sunil R. (2012). WHO classification of tumours of the breast: International Agency for Research on Cancer.

30. Nik-Zainal S, Davies H, Staaf J, Ramakrishna M, Glodzik D, Zou X, Martincorena I, Alexandrov LB, Martin S, Wedge DC *et al*. (2016). Landscape of somatic mutations in 560 breast cancer whole-genome sequences. *Nature*, **534**(7605): 47–54.

31. Jiang YZ, Ma D, Suo C, Shi J, Xue M, Hu X, Xiao Y, Yu KD, Liu YR, Yu Y *et al*. (2019). Genomic and transcriptomic landscape of triple-negative breast cancers: Subtypes and treatment strategies. *Cancer Cell*, **35**(3): 428–440 e425.

32. Abkevich V, Timms KM, Hennessy BT, Potter J, Carey MS, Meyer LA, Smith-McCune K, Broaddus R, Lu KH, Chen J *et al*. (2012). Patterns of genomic loss of heterozygosity predict homologous recombination repair defects in epithelial ovarian cancer. *Br J Cancer*, **107**(10): 1776–1782.

33. Birkbak NJ, Wang ZC, Kim JY, Eklund AC, Li Q, Tian R, Bowman-Colin C, Li Y, Greene-Colozzi A, Iglehart JD *et al*. (2012). Telomeric allelic imbalance indicates defective DNA repair and sensitivity to DNA-damaging agents. *Cancer Discov*, **2**(4): 366–375.

34. Popova T, Manie E, Rieunier G, Caux-Moncoutier V, Tirapo C, Dubois T, Delattre O, Sigal-Zafrani B, Bollet M, Longy M *et al*. (2012). Ploidy and large-scale genomic instability consistently identify basal-like breast carcinomas with BRCA1/2 inactivation. *Cancer Res*, **72**(21): 5454–5462.

35. von Minckwitz G, Schneeweiss A, Loibl S, Salat C, Denkert C, Rezai M, Blohmer JU, Jackisch C, Paepke S, Gerber B *et al*. (2014). Neoadjuvant carboplatin in patients with triple-negative and HER2-positive early breast cancer (GeparSixto; GBG 66): A randomised phase 2 trial. *Lancet Oncol*, **15**(7): 747–756.

36. Loibl S, Weber KE, Timms KM, Elkin EP, Hahnen E, Fasching PA, Lederer B, Denkert C, Schneeweiss A, Braun S *et al*. (2018). Survival analysis of carboplatin added to an anthracycline/taxane-based neoadjuvant chemotherapy and HRD score as predictor of response-final results from GeparSixto. *Ann Oncol*, **29**(12): 2341–2347.

37. Davies H, Glodzik D, Morganella S, Yates LR, Staaf J, Zou X, Ramakrishna M, Martin S, Boyault S, Sieuwerts AM *et al.* (2017). HRDetect is a predictor of BRCA1 and BRCA2 deficiency based on mutational signatures. *Nat Med*, **23**(4): 517–525.

38. Staaf J, Glodzik D, Bosch A, Vallon-Christersson J, Reutersward C, Hakkinen J, Degasperi A, Amarante TD, Saal LH, Hegardt C *et al.* (2019). Whole-genome sequencing of triple-negative breast cancers in a population-based clinical study. *Nat Med*, **25**(10): 1526–1533.

39. Shao F, Sun H, Deng CX. (2017). Potential therapeutic targets of triple-negative breast cancer based on its intrinsic subtype. *Oncotarget*, **8**(42): 73329–73344.

40. Gwe AS, Jun KS, Cheungyeul K, Joon J. (2016). Molecular classification of triple-negative breast cancer. *J Breast Cancer*, **19**(3): 223–230.

41. Lehmann BD, Bauer JA, Chen X, Sanders ME, Chakravarthy AB, Shyr Y, Pietenpol JA. (2011). Identification of human triple-negative breast cancer subtypes and preclinical models for selection of targeted therapies. *J Clin Invest*, **121**(7): 2750–2767.

42. Abramson VG, Lehmann BD, Ballinger TJ, Pietenpol JA. (2015). Subtyping of triple-negative breast cancer: Implications for therapy. *Cancer*, **121**(1): 8–16.

43. Lehmann BD, Jovanovic B, Chen X, Estrada MV, Johnson KN, Shyr Y, Moses HL, Sanders ME, Pietenpol JA. (2016). Refinement of triple-negative breast cancer molecular subtypes: Implications for neoadjuvant chemotherapy selection. *PloS one*, **11**(6): e0157368.

44. Masuda H, Baggerly KA, Wang Y, Zhang Y, Gonzalez-Angulo AM, Meric-Bernstam F, Valero V, Lehmann BD, Pietenpol JA, Hortobagyi GN *et al.* (2013). Differential response to neoadjuvant chemotherapy among 7 triple-negative breast cancer molecular subtypes. *Clin Cancer Res*, **19**(19): 5533–5540.

45. Burstein MD, Tsimelzon A, Poage GM, Covington KR, Contreras A, Fuqua SA, Savage MI, Osborne CK, Hilsenbeck SG, Chang JC. (2015). Comprehensive genomic analysis identifies novel subtypes and targets of triple-negative breast cancer. *Clin Cancer Res*, **21**(7): 1688–1698.

46. Liu YR, Jiang YZ, Xu XE, Yu KD, Jin X, Hu X, Zuo WJ, Hao S, Wu J, Liu GY. (2016). Comprehensive transcriptome analysis identifies novel molecular subtypes and subtype-specific RNAs of triple-negative breast cancer. *Breast Cancer Res*, **18**(1): 33.

47. Zhao S, Ma D, Xiao Y, Li XM, Ma JL, Zhang H, Xu XL, Lv H, Jiang WH, Yang WT *et al.* (2020). Molecular subtyping of triple-negative

breast cancers by immunohistochemistry: Molecular basis and clinical relevance. *Oncologist*.

48. Weigelt B, Kreike B, Reis-Filho JS. (2009). Metaplastic breast carcinomas are basal-like breast cancers: A genomic profiling analysis. *Breast Cancer Res Treat*, **117**(2): 273–280.

49. Prat A, Parker JS, Karginova O, Fan C, Livasy C, Herschkowitz JI, He X, Perou CM. (2010). Phenotypic and molecular characterization of the claudin-low intrinsic subtype of breast cancer. *Breast Cancer Res*, **12**(5): R68.

Chapter THREE

Genetics and Signaling Events in Triple-Negative Breast Cancer

Ying Li[1,2], Liang Zeng[3,*], *and* Faqing Tang[4,*]

Contents

*Corresponding authors: Liang Zeng, E-mail: zlxx03@126.com; Faqing Tang, E-mail: tangfaqing33@hotmail.com
[1]Key Laboratory of Translational Cancer Stem Cell Research, Hunan Normal University, Changsha, Hunan, China.
[2]Departments of Pathology and Pathophysiology, Hunan Normal University School of Medicine, Changsha, Hunan, China.
[3]Department of Pathology, Guangzhou Women and Children's Medical Center, Guangzhou Medical University, Guangzhou, Guangdong, China.
[4]Department of Laboratory Medicine, Hunan Cancer Hospital & The Affiliated Cancer Hospital of Xiangya School of Medicine, Central South University, Changsha, Hunan, China.

Genomic alterations occurring in either germline or somatic cells and changes in signaling molecules are crucial to cancer development and can also provide vulnerabilities to actionable drug targets. Triple-negative breast cancer (TNBC) is a type of breast cancer with poor prognosis. At present, the efficiency of conventional chemotherapy and targeted therapy is not ideal for TNBC patients. It is very important to find new therapeutic targets through elucidation of genetic and signaling changes. This chapter focuses on genomic alterations and the role of key signaling pathways in TNBC. The efforts that target these alterations will be the topic of Chapter 7.

3.1 Germline Mutations in TNBC

3.1.1 *BRCA1/2 Mutations and Homologous Recombination Defects*

BRCA1 and BRCA2 belong to a class of genes known as tumor suppressor genes [1]. By helping to repair DNA, BRCA proteins play a critical role in maintaining genomic stability and cell survival. In the cells that are deficient for BRCA1/2, DNA damages cannot be repaired by homologous recombination (HR). Rather, these damages are repaired by alternative repair pathways which are error-prone, leading to gross genomic instability or cell death (**Figure 3-1**).

The majority of BRCA1-mutated breast cancers are so-called "triple-negative" or of "basal-type". In contrast, only a minor proportion (<20%) of TNBC patients have a germline BRCA1/2 mutation. Germline mutations in BRCA1/2 genes are the most important cause of hereditary breast cancer [2]. The average

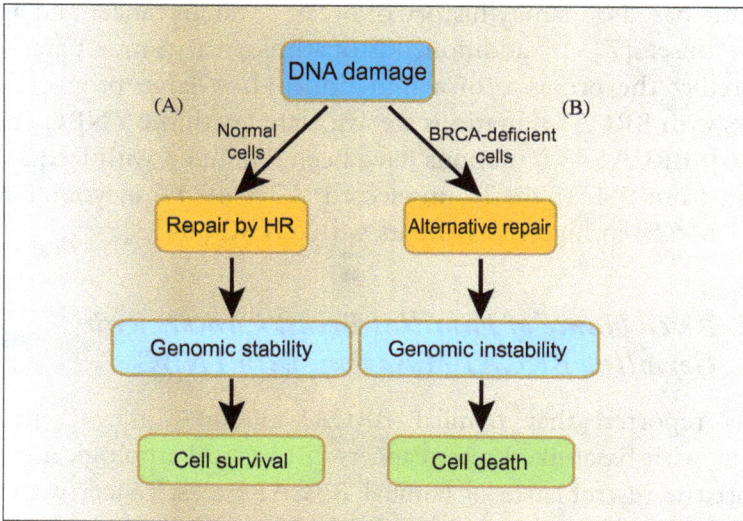

Figure 3-1. DNA repair defect and its effect on BRCA-deficient cells. (A) In normal cells, DNA damages are mostly repaired by homologous recombination (HR), which is dependent on functional BRCA1/2 proteins. (B) In BRCA-deficient cells, DNA damages cannot be repaired by HR, but instead, by potentially mitogenic alternative repair pathways, leading to gross genomic instability and/or cell death.

cumulative risk of breast cancer in female carriers above 70 years of age is estimated to be 57–65% and 45–49% for BRCA1 and BRCA2 mutation carriers, respectively [3]. Risk of hereditary cancers is assessed by taking into account familial and personal factors or clinicopathological characteristics of cancers such as TNBC [4]. With this information, women with a high risk of developing hereditary cancers are recommended to be tested for mutations in BRCA1/2 genes [5]. Women carrying a pathogenic germline mutation in BRCA1 and BRCA2 genes have an increased life-time risk of developing breast, ovarian, and several other cancers [6]. The identification of women harboring mutations in these genes is clinically important and has a significant socio-cultural impact. A major challenge faced by physicians is to identify most appropriate candidates for genetic BRCA1/2 testing since the cost of comprehensive genetic testing can be high and only 3% of all breast cancers are attributed to BRCA1/2 germline mutations.

The decision to offer genetic testing to a breast cancer patient is currently based on family history of breast/ovarian cancer and age of disease onset [7]. In addition, histopathological tumor parameters can predict the presence of a mutation [4]. A large proportion of tumors with BRCA1 mutations are associated with the TNBC phenotype [6]. BRCA1/2 mutations have been identified with frequencies varying from 9.4–15.4% in unselected, 17.4–49.1% in younger age and 11.6–62% in high risk patients with TNBC [8–13].

3.1.2 Resemblance of Familial Breast Cancers with Germline BRCA1 Mutations with TNBC

It was reported that familial BRCA1-mutant tumors segregate strongly with basal-like breast cancers (BLBCs). Pathological studies support the resemblance of familial BRCA1 breast cancer with sporadic BLBCs. This suggests that BRCA1 mutation might impose a defined gene expression pattern mandating basal characteristics in these tumors. It should be noted that familial BRCA2 tumors have features distinct from familial BRCA1 tumors, reflecting the functional difference between these two BRCA genes.

The similar features between familial BRCA tumors and BLBCs suggest that these tumors may share similar etiology, pathogenesis, progression, and may have special therapeutic implications for patients with these tumors.

3.1.2.1 *Association of BRCA1/2 germline mutations and tumorigenesis of TNBC*

Approximately only 15% of sporadic TNBCs are associated with germline mutations in BRCA1/2. However, in hereditary TNBCs, a high percentage of cases, approximately 73%, were BRCA1/2-associated [14]. From the reports from different countries and districts such as Germany, Japan, and China [15–17], the occurrence of TNBC is significantly associated with the BRCA1 mutation carrier status, and a different 'genetic background' may have a phenotypic impact on the onset of breast cancer.

3.1.2.2 *Association of BRCA1/2 germline mutations and progression & prognosis of TNBC*

In 194 cases of TNBC patients, 50 (26%) germline mutation carriers (78% in BRCA1) and 136 (71%) tumors with somatic mutations (83% in TP53) were reported. Tumor mutation patterns differed between carriers and non-carriers. PIK3CA mutations were exclusively present in non-carriers. Tumors that had lost the germline mutation demonstrated a higher incidence of somatic TP53 mutations as compared to tumors with preserved germline mutations. Germline mutation status significantly interacted with tumor TP53 mutations were used for patient disease-free survival. In non-carriers, TP53 mutations did not affect outcome; in carriers, those with mutated TP53 tumors experienced more relapses compared to those with wild-type TP53 tumors. The tumor genotypes with respect to germline status and the prognostic interaction between germline BRCA1-related and tumor TP53 mutation status prompt combination of germline and tumor genotyping for the classification of TNBC, particularly in the context of clinical trials evaluating synthetic lethality drugs [18].

Effects of BRCA germline mutations on multiple survival outcomes of breast cancer patients were evaluated in specific subgroups, including the patients with TNBC. When all subtypes are considered, BRCA1 mutation carriers have worse overall survival (OS) and worse breast cancer-specific survival than sporadic/BRCA-wild-type breast cancer cases. However, among TNBC, BRCA1/2 mutation carriers have better OS than BRCA-wild-type counterpart.

The prognosis of BRCA1/2-associated breast cancer partly depends on histologic characteristics. Independent prognostic factors include tumor size, tumor-associated inflammation, and intratumor necrosis. Established prognostic factors as nodal status and differentiation grade were not significantly related to relapse-free survival (RFS). Tumor-associated inflammation density was the strongest predictor for RFS in this series of BRCA1/2 breast cancer patients [19]. Bayraktar *et al.* revealed 50% prevalence of deleterious BRCA1/2 mutations in high-risk women diagnosed with TNBC. Overall prognosis of TNBC in BRCA carriers and non-carriers is not significantly different within the first 5 years following initial diagnosis [20]. Afghahi *et al.* reported that a new BRCA1 mutation was detected in the residual disease which resulted in a 14-amino acid deletion and restoration of the BRCA1 reading frame. A local relapse biopsy 4 months later revealed the identical reversion mutation, and the patient subsequently died from metastatic breast cancer [21].

3.1.2.3 *Association of BRCA1/2 germline mutations and therapy of TNBC*

Traditionally, BRCA carriers have received conventional systemic chemotherapy based on their baseline tumor characteristics, and it is generally accepted that after the appropriate treatment the prognosis of a mutation carrier is equivalent to that of a patient with sporadic breast cancer. With the growing understanding of the functions of BRCA1/2 proteins in homologous DNA repair, it is recognized that BRCA-associated breast tumors may have distinct biochemical characteristics and thus require tailored treatment strategies. Tumors arising in the patients with BRCA mutations were shown to be particularly sensitive to poly (ADP-ribose) polymerase inhibitors (PARPi's) (discussed in Chapter 6) or platinum compounds (discussed in Chapter 7). In addition, BRCA1-mutation carriers seem to benefit from anthracycline-taxane-containing regimens as much as sporadic TNBCs do [22].

Recent advancements in subclassifying TNBC have paved the way for further investigation of more effective systemic therapies, including cytotoxic and targeted agents. TNBC is enriched for germline BRCA mutation and for somatic deficiencies in HR-mediated DNA repair, the so-called "BRCAness" phenotype [1]. Together, germline BRCA mutations and BRCAness are promising biomarkers of susceptibility to DNA-damaging therapy.

3.2 Somatic Gene Mutations in TNBC

Cancer genomes harbor a large number of somatic alterations, but only a few of them play a significant role in driving carcinogenesis by conferring selective advantage to tumor cell growth [23]. Genomic profiling studies of human TNBC, including whole-exome and whole-genome analyses, have identified numerous recurrent alterations in these cancer driver genes. Among them, somatic mutations and copy number alterations (CNAs) account for the most commonly encountered genomic alterations in TNBC. TP53 is the most frequently mutated gene, but at present no drugs targeting TP53 have been approved for clinical practice. Aside from TP53, a handful of other genes with >5% prevalence of mutation have been identified in TNBC (discussed below). MYC amplification is the most frequent CNA event in TNBC. Other genes frequently affected by somatic CNAs include EGFR, PTEN, CCND1, RB1, and CCNE1 [24].

3.2.1 *Somatic Gene Mutations and Tumorigenesis of TNBC*

Rechsteiner *et al.* analyzed 32 clinically sporadic breast cancers with medullary features for somatic BRCA1/2 mutations. 3 of 32 tumors had pathogenic BRCA1 gene alterations. Two of these pathogenic variants exhibited deletions leading to frame shift mutations (p.Glu23fs, p.Val1234fs), and the remaining single-nucleotide variant resulted in premature STOP codon (p.Glu60Ter). In one patient, the same pathogenic BRCA1 mutation was detected (p.Glu23fs) in normal breast tissue [25].

KDR, PIK3CA, Akt1, ATM, BRCA1/2, TP53, and KIT were among the most frequently mutated genes in TNBC cohort (**Figure 3-2**). The SNP Akt1 (rs3730358) was suggested to modify the risk of breast cancer. SNP PIK3CA (rs3729687) is a damaging mutation that was found to be correlated with the prognosis of TNBC. The survival curve analysis showed that the presence of Akt1, TP53, KDR, KIT, BRCA1, and BRCA2 mutations is correlated with a poor prognosis [26].

In 194 TNBC patients, 50 (26%) germline mutation carriers (78% in BRCA1) and 136 (71%) tumors with somatic mutations (83% in TP53) were identified. PIK3CA mutations were exclusively present in non-carriers. The germline mutation demonstrated a higher incidence of somatic TP53 mutations as compared to tumors with preserved germline mutations. In non-carriers, tumor TP53 mutations did not affect outcome. In carriers, those with mutated TP53 tumors experienced more relapses compared to those with wild-type TP53 tumors [18]. Intratumor heterogeneity (ITH) plays a pivotal role in driving breast cancer progression and therapeutic resistance. High T stage,

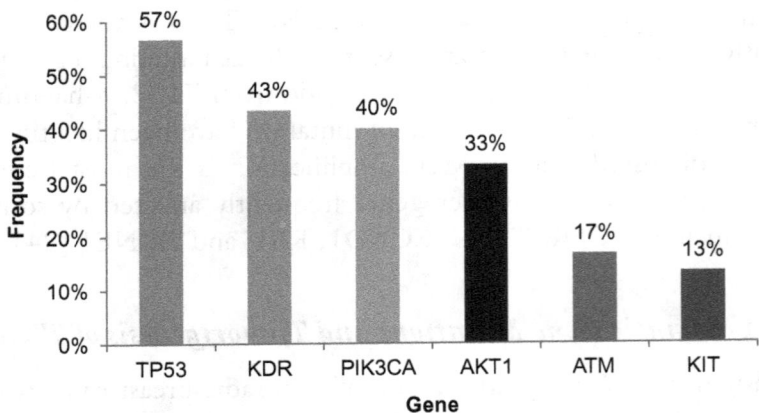

Figure 3-2. Frequency of mutations in specific genes in TNBC. Ref: *The Breast. 2018; 38:30–38.*

African American race, and triple-negative or basal-like subtype were associated with a higher mutant-allele tumor heterogeneity (MATH) level [27].

It was reported that 18/20 of TNBCs contained at least one detected somatic mutation. TP53, AURKA, and KDR mutations were each present in 6/20 of cases [28]. The two most common alterations in breast cancer are TP53 affecting the majority of TNBC and PIK3CA mutations affecting almost half of ER-positive cancers [29]. Identical TP53 mutations and similar patterns of gene CNAs were found in MGA and/or AMGA and in the associated TNBC. In the MGA/atypical MGA associated with TNBC lacking TP53 mutations, somatic mutations affecting PI3K pathway-related genes (e.g., PTEN, PIK3CA and INPP4B) and tyrosine kinase receptor signaling-related genes (e.g., ErbB3 and FGFR2) were identified. The heterogeneity of MGAs are associated with TNBC, and MGAs are genetically advanced, clonal, and neoplastic lesions harboring recurrent mutations in TP53 and/or other tumor genes, supporting the notion that a subset of MGAs and AMGAs may constitute non-obligate precursors of TNBCs [30].

Somatic mutations in the PI3K p110 catalytic subunit (PIK3CA) gene are common. Activating mutations in PIK3CA have been identified in 18–40% of breast carcinomas. PIK3CA mutations were observed in 43/185 of breast tumor samples. PIK3CA mutations were common in ER⁺, PR⁺ and HER2⁺ cases (30%), and a comparatively low frequency were noted in triple-negative tumors (13.6%) [31]. TNBCs with apocrine differentiation less frequently harbored TP53 mutations (25%) and displayed a high mutation frequency in PIK3CA and other PI3K signaling pathway-related genes (75%). A high frequency of PI3K pathway alterations in TNBC was similar to luminal subtypes of breast cancer [32].

Although somatic mutations in BRCA1 rarely occur in sporadic breast cancer, lower than normal rates of expression of BRCA1 is reported to be an important factor that contributes to tumorigenesis in sporadic tumors. The epigenetic inactivation of BRCA1 expression might thus play an important role in sporadic breast cancer cases. BRCA1 promoter methylation was found in 11 tumors and all of these were in 69 TNBC cases and was significantly associated with lymphovessel invasion, high nuclear grade, low BRCA1 mRNA expression, loss of BRCA1 protein expression and shorter OS [33].

The MET receptor tyrosine kinase is elevated in TNBC and transgenic Met models develop basal-like tumors. Somatic Trp53 loss, in combination with Met abnormality, significantly increased tumor penetrance over Met abnormality or Trp53 loss alone. The majority of Met tumors with Trp53 loss displayed spindloid pathology with distinct molecular signature that resembles the human claudin-low subtype of TNBC, including diminished claudins, EMT signature, and low expression of microRNA-200 family. Among human breast cancers, elevated levels of MET and stabilized TP53, indicative of mutation, correlate with highly proliferative TNBCs of poor outcome [34].

Id4, a helix-loop-helix DNA binding factor, blocks BRCA1 gene transcription in vitro and downregulates BRCA1 in vivo [35]. Although TP53, PIK3CA and PTEN somatic mutations seem to be clonally dominant compared to other genes, in some tumors their clonal frequencies are incompatible with founder status. Mutations in cytoskeletal, cell shape and motility proteins occurred at lower clonal frequencies, suggesting that they occurred later during tumor progression [36]. KRAS mutations are extremely infrequent in TNBC and EGFR inhibitors may be of potential benefit in the treatment of BLBCs, which overexpress EGFR in about 60% of all cases [37].

3.2.2 *Somatic Gene Mutations and Progression & Prognosis of TNBC*

Chromosome 5q loss is detected in up to 70% of TNBCs. Somatic deletion of a region syntenic with human 5q 33.2–35.3 was showed in a mouse model of TNBC. Mechanistically, KIBRA as a major factor contributing to the effects of 5q loss on tumor growth and metastatic progression was identified. Reexpression of KIBRA impairs cancer metastasis in vivo and inhibits tumor sphere formation of TNBC cells in vitro. 5q loss involves the reduced dosage of KIBRA, promoting oncogenic functioning of YAP/TAZ in TNBC [38]. $CD4^+$ and HLA-DRB1 * 1501-restricted TILs isolated from breast cancer were recognized to have a single mutation in RBPJ (recombination signal binding protein for immunoglobulin kappa J region). Analysis of 16 metastatic sites revealed that the mutation was ubiquitously present in all samples. Breast cancers can express naturally processed and presented unique nonsynonymous mutations that are recognized by a patient's immune system. TILs recognizing these immunogenic mutations can be isolated from a patient's tumor, suggesting that adoptive cell transfer of mutation-reactive TILs could be a viable treatment option for patients with breast cancer [39].

It was reported that the TP53 mutation frequency was higher in brain metastasis than in primary breast cancer [40]. BRCA1/2 mutations were associated with bilateral breast cancer, and BRCA1 promoter methylation may have a prognostic effect on TNBC [41]. Most germline pathogenic variants occurred in BRCA1 gene. BRCA1 promoter hypermethylation was detected in 20.6% of tumors, and none of these tumors were in BRCA1/2 pathogenic variant carriers. BRCA1 impairment by either germline or somatic events was significantly more frequent in young women (55% in those ≤40 years of age; 33% in those between 41–50 years of age; 22% in those >50 years of age), and associated with better OS and DFS rates in breast cancer patients [42].

Somatic mutation of p53 is rare, suggesting that p53 becomes inactivated by other mechanisms. There are several important clinical associations, particularly with Δ40p53, which was expressed at levels that were ~50-fold higher than the least expressed isoform p53γ. Δ40p53 was significantly upregulated in tumor tissue when compared with the normal breast, and was significantly associated with an aggressive TNBC. Additionally, p53β expression was significantly negatively associated with tumor size, and positively associated with DFS, where high levels of p53β were protective, particularly in the patients with a mutation in p53, suggesting p53β may counteract the damage inflicted by mutant p53 [43].

Out of 507 breast cancer patients enrolled in West China Hospital, 9 patients had somatic variants (3 in BRCA1, 6 in BRCA2) and 1 patient had concurrent germline/somatic variants in BRCA2. In patients with disease stage 0–II, presence of germline or somatic

BRCA1 P/LP variant increased the risk of relapse as compared to non-carriers. Germline BRCA1 P/LP variants, which were associated with aggressive tumor phenotypes, predicted worse DFS in the subgroup of stage 0–II compared to non-carriers [44]. In 77 TNBC and normal tissues, 15 patients (19.5%) had BRCA mutations: 12 (15.6%) in BRCA1 (one somatic), and 3 (3.9%) in BRCA2. Patients with BRCA mutations tended to be younger and had a significantly better RFS than those with wild-type BRCA [45].

3.2.3 *Somatic Gene Mutations and Therapy of TNBC*

Somatic mutations of PIK3CA occur with high frequency in patients. In a subset of TNBC cell lines, treatment with a PI3K inhibitor or depletion of PIK3CA expression ultimately promoted Akt reactivation in a manner dependent on the E3 ubiquitin ligase Skp2, the kinases IGF-1R and PDK-1 (phosphoinositide-dependent kinase-1), and the cell growth and metabolism-regulating complex mTORC2 (mammalian target of rapamycin complex 2), but was independent of PI3K activity or PIP3 production. Resistance to PI3Ki's are correlated with the increased abundance of Skp2, ubiquitination of Akt, cell proliferation in culture and xenograft tumor growth in mice [46].

Immune checkpoint inhibitors have emerged as a potent new class of anti-cancer therapy (discussed in Chapter 8). Interestingly, immune checkpoint inhibition is more effective in those tumors with a high mutational load. In general, modest results have been observed in breast cancer, where tumors are rarely hypermutated. Because BRCA1-muated tumors frequently exhibit a triple-negative phenotype with extensive lymphocyte infiltration, the mutational load, immune profile, and response to checkpoint inhibition were explored in a BRCA1-deficient tumor model. BRCA1-mutated TNBCs exhibited an increased somatic mutational load and great numbers of TILs, with increased expression of immunomodulatory genes including PD1 and CTLA4, when compared to TNBCs from BRCA1-wild-type patients.

Cisplatin treatment combined with dual anti-PD1 and anti-CTLA4 therapy substantially augmented anti-tumor immunity in

BRCA1-deficient mice, resulting in an avid systemic and intratumoral immune response. This response involved the enhanced dendritic cell activation, reduced suppressive FOXP3⁺ regulatory T cells, and con-comitant increased the activation of tumor-infiltrating CD8⁺ and CD4⁺ T cells, characterized by the induction of polyfunctional cytokine-producing T cells. Dual but not single checkpoint blockade together with cisplatin profoundly attenuated the growth of BRCA1-deficient tumors in vivo and improved survival [47].

3.3 Signaling Events in TNBC

Due to the complexity of the genomic landscape, analysis of altera-tions in single genes is often insufficient for tumor classification. Thus, effort has been made to group the aberrations of individual genes according to the molecular pathways. Such grouping has helped us better understand the tumor biology and facilitated the development of drugs targeting these pathways. The detailed altera-tions in prominent signaling pathways in TNBC are discussed here.

3.3.1 *PI3K/Akt/mTOR Pathway*

The PI3K/Akt/mTOR pathway is a prototypic survival pathway that is constitutively activated in many types of cancer (**Figure 3-3**) and this pathway is an attractive therapeutic target in cancer because it serves as a convergence point for many growth stimuli, and through its downstream substrates which controls cellular processes contribut-ing to the initiation and maintenance of cancer. Moreover, activation of the Akt/mTOR pathway confers resistance to many types of cancer therapy, and is a poor prognostic factor for many types of cancers. Combined mutations of tumor suppressors PTEN and p53 accelerate the formation of claudin-low breast cancers and TNBC exhibits hyperactivated Akt signaling and more mesenchymal features relative to PTEN or p53 single-mutant tumors [48]. PIK3CA mutations were detected in 23.7% of TNBC. Deregulation of PI3K/Akt pathways was revealed by consistent activation of pAkt and p-p44/42 MAPK in all PIK3CA-mutated TNBC [49].

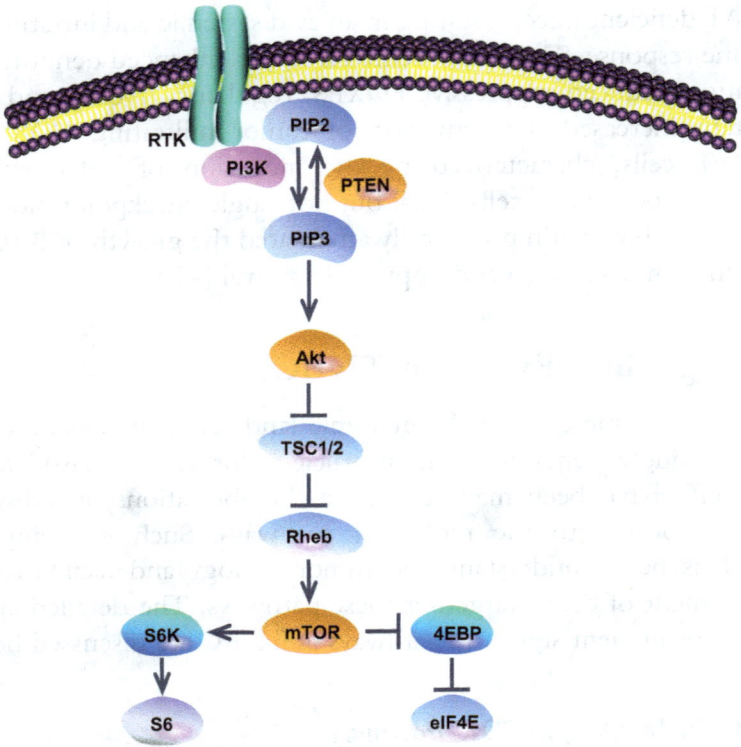

Figure 3-3. PI3K/Akt/mTOR signaling pathway. When activated, for example, by RTK, PI3K converts PIP2 to PIP3 through phosphorylation. PIP3 leads to the activation of Akt, which induces stepwise phosphorylation events, ultimately leading to the activation of mTOR, which plays a central role in the translational machinery through 4EBP and S6K pathways. The final outcome of the activation of the PI3K/Akt/mTOR singling pathway is the regulation of various cellular functions including cell growth and survival related to oncogenic phenotypes. As a negative regulator, PTEN converts PIP3 back to PIP2, thus dampening the PI3K/Akt/mTOR signaling activity. RTK, receptor tyrosine kinase; PIP2, phosphatidylinositol (4,5)P2; PIP3, phosphatidylinositol (3,4,5)P3; TSC1/2, tuberous sclerosis complex 1/2; Rheb, RAS homologue enriched in brain; mTOR, mammalian target of rapamycin; S6, ribosomal S6 protein; S6K, ribosomal S6 protein kinase; eIF4E, eukaryotic initiation factor 4E; 4EBP, eIF4E-binding protein.

3.3.2 Wnt Pathway

3.3.2.1 Association of Wnt pathway and tumorigenesis of TNBC

Wnt signaling pathways are a group of signal transduction pathways made of proteins that pass signals into a cell through cell surface receptors. Three Wnt signaling pathways have been characterized: canonical Wnt/β-catenin pathway, non-canonical Wnt planar cell polarity pathway, and non-canonical Wnt/calcium pathway. Wnt signaling was first identified for its role in carcinogenesis, then for its function in embryonic development. In the embryonic processes, it controls body axis patterning, cell fate specification, cell proliferation, and cell migration [50] (**Figure 3-4**).

In TNBC, the elevated levels of transforming growth-interacting factor (TGIF) correlate with high Wnt signaling and poor survival of patients. Genetic experiments revealed that TGIF1 ablation impeded mammary tumor development in MMTV-Wnt1 mice [51]. Wnt signaling appears to be active in both the normal and cancer stem cell compartments, although at different levels. By comparing normal with cancer mouse mammary compartments, mammary cancer stem cells (MaCSCs) gene signature was identified to be able to predict outcome in breast cancer. Wnt signaling activation affects self-renewal and differentiation of MaCSCs, leading to metaplasia and basal-like adenocarcinomas [52].

WIF-1 is a secreted antagonist that binds Wnt ligands, and therefore inhibits the canonical Wnt/β-catenin pathway. Methylation of WIF-1 was detected in 71.3% and 81.8% of sporadic and hereditary breast cancer cases, respectively. Aberrant methylation of WIF-1 was associated with advanced TNM stage and triple-negative cases in sporadic breast carcinoma. In hereditary cases, methylation of WIF-1 was correlated with age at diagnosis and the p53 status. Regarding patients' survival, WIF-1 methylated promoter conferred a reduced OS rate, and particularly in a group of patients with advanced TNM stage. Aberrant CpG methylation of WIF-1 promoter was significantly associated with transcriptional silencing of this tumor suppressor gene in sporadic breast cancer tissues [53].

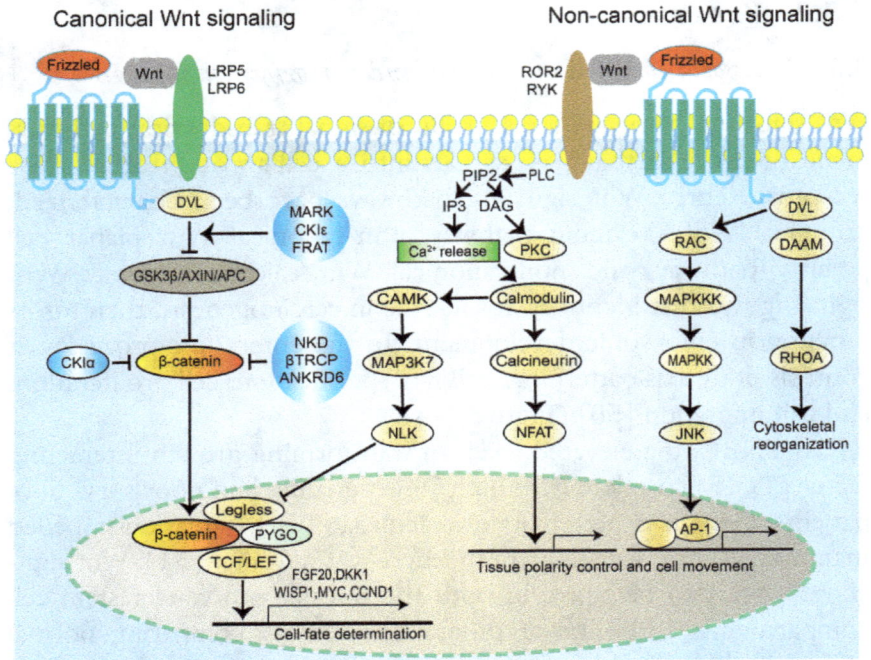

Figure 3-4. Canonical and non-canonical Wnt signaling pathways. Wnt signals are transduced to the canonical pathway for cell-fate determination and to the non-canonical pathway for the control of tissue polarity and cell movement. Activation of the canonical Wnt/β-catenin pathway promotes the accumulation of β-catenin by inhibiting the formation of destruction complex (GSK3β/AXIN/APC), resulting in the translocation of β-catenin to the nucleus and the expression of cell-fate determination genes. Activation of the non-canonical Wnt signaling pathway promotes the expression of genes involved in tissue polarity control and cell movement. LRP, LDL receptor related protein; DVL, dishevelled; PKC, protein kinase C; RYK, receptor tyrosine-like kinase; TCF/LEF, T-cell factor/lymphoid enhancer factor; CAMK, calcium calmodulin-dependent protein kinase; DAAM, disheveled-associated activator of morphogenesis; NFAT, nuclear factor of activated T-cells.

Wnt/β-catenin signaling may play a critical role in breast cancer immunity, particularly in HER2-enriched subtype and TNBC. Both stromal infiltrated TILs and β-catenin expression were upregulated in hormone receptor-negative HER2-enriched and TNBC subtypes, and high levels of stromal TILs as well as CD8[+] or FOXP3[+] TIL subsets were associated with β-catenin overexpression by breast cancer, respectively [54].

Analysis of publically available array data sets indicates that the tumors with concomitant low expression of Wnt-regulating proteins APC (adenomatous polyposis coli) and APC2 occurs more frequently in the 'triple-negative' phenotype. Upon aging, the development of Wnt-activated mammary carcinomas with squamous differentiation was accompanied by a significantly reduced survival. This novel Wnt-driven mammary tumor model highlights the importance of functional redundancies existing between APC proteins both in normal homeostasis and in tumorigenesis [55].

Diversin was reported to play roles in Wnt and JNK pathways. Significant association was observed between diversin overexpression and TNM stage, nodal metastasis, negative ER expression and triple-negative status. Knockdown of diversin expression in MDA-MB-231 cell line decreased cell proliferation and cell invasion [56]. Knockdown of Wnt pathway transcription factor, SOX4 in triple-negative BT-549 cells, resulted in decreasing cell proliferation and migration. The combination treatment of Wnt pathway inhibitors iCRT-3 with SOX4 knockdown had a synergistic effect on inhibition of cell proliferation and induction of apoptosis in TNBC cells [57].

Wnt10B, a member of the 19 Wnt family ligands, which activates the canonical Wnt signaling cascade, induces transcriptionally active β-catenin in human TNBC, and predicts survival outcome of patients with both TNBC and BL tumors. Wnt10B activates canonical β-catenin signaling leading to upregulation of HMGA2, which is necessary and sufficient for proliferation of TNBC cells and predicts RFS and metastasis in TNBC patients [58].

3.3.2.2 Association of Wnt pathway and progression & prognosis of TNBC

Cancer stem cells (CSCs) are believed to promote the malignant transformation of breast cancer, at least partly, via Wnt/β-catenin pathway. Leucine-rich repeat-containing G protein-coupled receptor 5 (LGR5) has been identified as a CSC-associated Wnt-regulated target gene. High levels of LGR5 expression were significantly associated with tumor size, lymph node metastasis, and TNBC. Patients with high levels of LGR5-β-catenin axis expression exhibited poorer RFS compared to patients with low levels of LGR5-β-catenin axis expression [59]. MCC (mutated in colorectal cancers) and CTNNBIP1 (β-catenin-interacting protein 1) are two candidate genes which inhibit the transcriptional activity of nuclear β-catenin. The expression of nuclear p-β-catenin (Y654) was significantly high in TNBC and HER2$^+$ compared to luminal A/B subtypes. TNBC patients in stage III/IV had a reduced expression of MCC in the tumors with poor prognosis [60]. A statistically significant interaction was shown such that low expression of β-catenin in cell membrane was associated with unfavorable DFS of the tumors that expressed EGFR, but not in the absence of EGFR expression. A considerable number of TNBC co-expresses E-cadherin and P-cadherin, while membranous localization of β-catenin may predict patient outcome in an EGFR-dependent manner [61].

In vivo and in vitro data uncovered that Wnt5B-modulated mito-chondrial physiology was mediated by MCL1, which was regulated by Wnt/β-catenin responsive gene, Myc. Clinic data revealed that both Wnt5B and MCL1 are associated with enhanced metastasis and decreased DFS [62]. Expression of dickkopf-1 (DKK-1) in TNBCs is correlated with cytoplasmic/nuclear β-catenin. Elevated expression of DKK1 and cytoplasmic/nuclear β-catenin indicate poor outcome of TNBC patients [63].

3.3.3 *MAPK Pathway*

The MAPK/ERK pathway consists of a chain of proteins in the cell that communicates a signal from a receptor on cell surface to nuclear DNA. The signal starts when a signaling molecule binds to the receptor on the cell surface, and ends when DNA in the nucleus expresses a protein and produces some changes in the cell, such as cell division, survival, and/or motility/invasion (**Figure 3-5**). The pathway includes many proteins, which communicate by adding phosphate groups to a neighboring protein, acts as an "on" or "off" switch. When one of the proteins in the pathway is mutated, it can become stuck in the "on" or "off" position, which is a necessary step in the

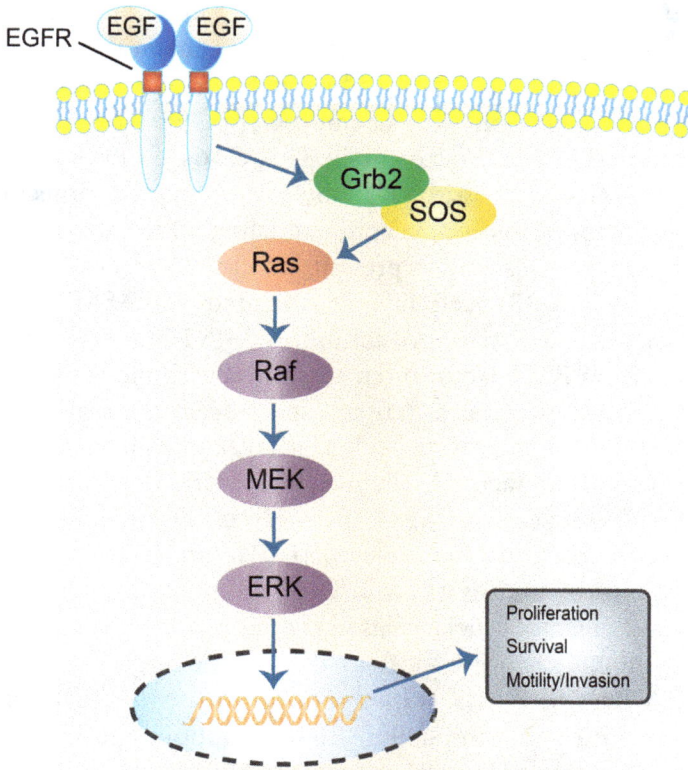

Figure 3-5. MAPK/ERK signaling pathway. EGFR forms an activated homodimer/heterodimer after ligand binding and is auto-phosphorylated at intrinsic tyrosine residues. Then, the phosphorylated EGFR induces Grb2/SOS complex formation and elicits cascaded activation of downstream proteins, such as Ras, Raf, MEK, and ERK. The signal is ultimately transmitted to the nucleus and initiates gene transcription for proliferation, survival, and motility/invasion. EGFR, epidermal growth factor receptor; Grb2, growth factor receptor-bound protein 2; MEK, mitogen-activated protein kinase kinase; ERK, extracellular signal-regulated kinase; SOS, son of sevenless homolog; Ras, rat sarcoma viral oncogene homolog; Raf, v-raf-1 murine leukemia viral oncogene homolog.

development of many cancers. Components of the MAPK/ERK pathway were discovered when they were found in cancer cells. Drugs that reverse the "on" or "off" switch are being investigated as cancer treatments.

Hypoxia-inducible factor (HIF)-1-dependent regulation of MAPK signaling pathways contributes to chemotherapy-induced breast CSC enrichment. Therapeutic targeting of HIF-1 or the p38 pathway in combination with chemotherapy will block breast CSC enrichment and improve outcome in TNBC [64]. Inhibitor of apoptosis proteins (IAPs) constitutes a family of conserved molecules that regulate both apoptosis and receptor signaling which are often deregulated in cancer cells and represent potential targets for therapy. TNBC MDA-MB-231 cell line was treated with SM83, a Smac mimetic that acts as a pan-IAP inhibitor. SM83 reduced the expression of Snai2, an EMT factor often associated with increased stem-like properties and metastatic potential especially in breast cancer cells. Snai2 downregulation prevents cell motility, and its expression is promoted by cIAP1. In fact, the chemical or genetic inhibition of cIAP1 blocked EGFR-dependent MAPK pathway activation, and caused the reduction of Snai2 transcription levels. IAP inhibition displays an anti-tumor and anti-metastasis effect in vivo [65].

Raf and ERK play crucial roles in the apoptosis resistance of breast cancer cells and are both important therapeutic targets in the MAPK pathway. A Raf/ERK dual inhibitor, CY-9d, was found to suppress breast cancer growth, inhibit Raf/ERK activation, and induce mitochondrial apoptosis in vivo without remarkable toxicity [66]. Cytotoxic and apoptosis-inducing activities of CY-9d were found to be restricted to TNBC cells. HSP90 was found to be a potential mediator between Raf and ERK in TNBC cells. Simultaneous treatment with HSP90 inhibitor and CY-9d at sub-therapeutic doses was found to produce synergistic therapeutic and apoptosis-inducing effects in TNBC cells. Aberrant Ras-MAPK signaling from RTKs, including EGFR and HER2, is a hallmark of TNBC. Raf and MEK co-inhibition exhibits synergy in TNBC models, and represent a promising combination therapy for this aggressive breast cancer type [67].

As discussed in Chapter 8, the presence of TILs in TNBC is correlated with improved outcomes, particularly in response to immune checkpoint-targeted immunotherapy. Ras/MAPK pathway activation is associated with significantly low levels of TILs in TNBC and while MEK inhibition can promote recruitment of TILs to the tumor.

α-4-1BB and α-OX-40 T-cell agonist antibodies can rescue the adverse effects of MEK inhibition on T cells in both mouse and human T cells, which results in augmented anti-tumor effects in vivo. This effect is dependent upon the increased p38/JNK pathway activation. MEK inhibition in breast cancer is associated with increased TILs, MAPK activity is required for T cells function. It shows that TILs activity following MEK inhibition can be enhanced by agonist immunotherapy, resulting in synergic therapeutic effects [68].

MEK pathway is activated in TNBC brain metastases, brain-penetrant inhibitors against MEK1/2 (selumetinib, AZD6244) or phosphatidylinositol-3 kinase (PI3K; buparlisib, BKM120) inhibit MEK pathway activation. Exploratory analysis of kinome reprogramming in SUM149 intracranial tumors after MEK and PI3K inhibition demonstrates extensive kinome changes with treatment, especially in MAPK pathway members. Rational combinations of the clinically available inhibitors selumetinib with buparlisib or pazopanib may prove to be promising therapeutic strategies for TNBC brain metastases [69].

The use of selenium-containing heterocyclic compounds as potent cancer chemopreventive and chemotherapeutic agents has well been documented by a large number of clinical studies. Liang *et al.* demonstrated that benzimidazole-containing selenadiazole derivatives (BSeDs) could cause cell cycle arrest and apoptosis in MDA-MB-231 cells by inducing DNA damage, inhibiting Akt, and activating MAPK family members through the production of reactive oxygen species (ROS) [70].

3.3.4 *Hedgehog Pathway*

Mammals have three Hedgehog homologues, Desert (DHH), Indian (IHH), and Sonic (SHH), of which Sonic is the best studied. The pathway is equally important during vertebrate embryonic development. Recent studies point to the role of Hedgehog signaling in regulating adult stem cells involved in maintenance and regeneration of adult tissues. The pathway has also been implicated in the development of some cancers. Dysregulation of the Hedgehog signaling pathway has been documented in mammary gland development and breast cancer progression. High levels of SHH was observed in a sub-set of breast tumors with poor prognostic pathological features; higher level of SHH expression was correlated with a significantly poor OS of patients. These data suggest that SHH could be a novel biomarker mediating the aggressive phenotype of breast cancer [71]. The major components and the signaling events in the Hedgehog signaling pathway are illustrated in **Figure 3-6**.

Hedgehog activation has been implicated in breast cancer poor prognosis. It was reported that 36 breast cancer patients had varying degrees of cytoplasmic sonic Hedgehog and its downstream effector glioma-associated oncogene homolog (Gli)-1 staining, which was

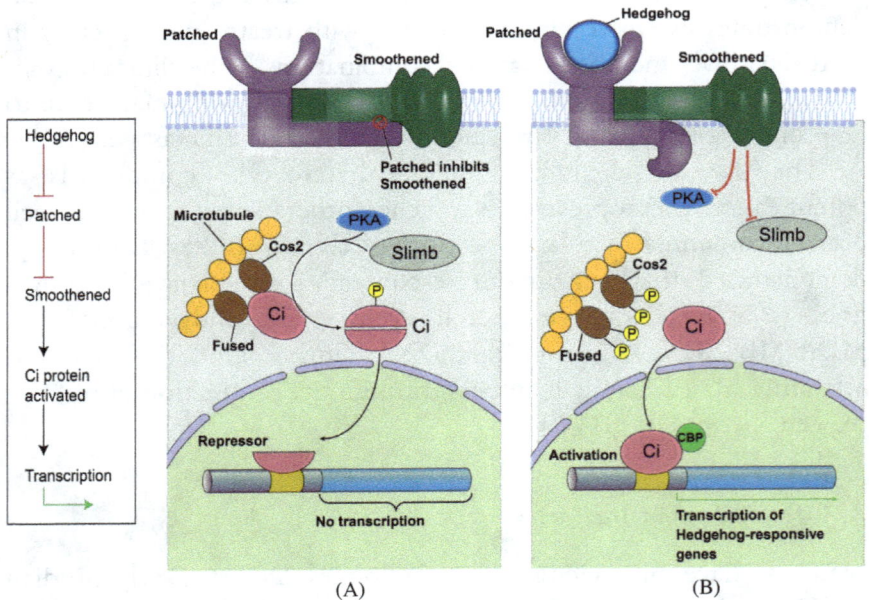

(A)   (B)

Figure 3-6.   Hedgehog signaling pathway. (A) Patched (Ptc) inhibits the activity of the Smoothened (Smo) protein, thereby inhibiting the downstream signaling pathway. At this time, the downstream Ci protein is truncated in the protease and enters the nucleus in the form of a truncated carboxyl terminus, which inhibits the transcription of the downstream target genes. (B) When Ptc and Hedgehog (Hh) are combined, the inhibitory effect on Smo is released, so that the full-length Ci protein gets activated, enters the nucleus, and activates transcription of the downstream target genes.

positively correlated with tumor stage [72]. Key regulators of the Hedgehog pathway components were significantly overexpressed in breast cancer tissues as compared with respective normal mammary tissues, overexpression of SHH, DHH and GLI1 were significantly related to the patients with early onset and pre-menopausal status [73].

References

1. Turner N, Tutt A, Ashworth A. (2004). Hallmarks of 'BRCAness' in sporadic cancers. *Nat Rev Cancer*, 4(10): 814–819.
2. Antoniou A, Pharoah PD, Narod S, Risch HA, Eyfjord JE, Hopper JL, Loman N, Olsson H, Johannsson O, Borg A *et al.* (2003). Average risks of breast and ovarian cancer associated with BRCA1 or BRCA2 mutations detected in case series unselected for family history: A combined analysis of 22 studies. *Am J Hum Genet*, 72(5): 1117–1130.
3. Chen S, Parmigiani G. (2007). Meta-analysis of BRCA1 and BRCA2 penetrance. *J Clin Oncol*, 25(11): 1329–1333.
4. Foulkes WD, Smith IE, Reis-Filho JS. (2010). Triple-negative breast cancer. *N Engl J Med*, 363(20): 1938–1948.
5. Moyer VA. (2014). Risk assessment, genetic counseling, and genetic testing for BRCA-related cancer in women: U.S. Preventive Services Task Force recommendation statement. *Ann Intern Med*, 160(4): 271–281.
6. Ford D, Easton DF, Stratton M, Narod S, Goldgar D, Devilee P, Bishop DT, Weber B, Lenoir G, Chang-Claude J *et al.* (1998). Genetic heterogeneity and penetrance analysis of the BRCA1 and BRCA2 genes in breast cancer families. The Breast Cancer Linkage Consortium. *Am J Hum Genet*, 62(3): 676–689.
7. Antoniou AC, Easton DF. (2006). Risk prediction models for familial breast cancer. *Future Oncol*, 2(2): 257–274.
8. Couch FJ, Hart SN, Sharma P, Toland AE, Wang X, Miron P, Olson JE, Godwin AK, Pankratz VS, Olswold C *et al.* (2015). Inherited mutations in 17 breast cancer susceptibility genes among a large triple-negative breast cancer cohort unselected for family history of breast cancer. *J Clin Oncol*, 33(4): 304–311.
9. Evans DG, Howell A, Ward D, Lalloo F, Jones JL, Eccles DM. (2011). Prevalence of BRCA1 and BRCA2 mutations in triple negative breast cancer. *J Med Genet*, 48(8): 520–522.

10. Hartman AR, Kaldate RR, Sailer LM, Painter L, Grier CE, Endsley RR, Griffin M, Hamilton SA, Frye CA, Silberman MA *et al.* (2012). Prevalence of BRCA mutations in an unselected population of triple-negative breast cancer. *Cancer*, **118**(11): 2787–2795.

11. Meyer P, Landgraf K, Hogel B, Eiermann W, Ataseven B. (2012). BRCA2 mutations and triple-negative breast cancer. *PloS One*, **7**(5): e38361.

12. Young SR, Pilarski RT, Donenberg T, Shapiro C, Hammond LS, Miller J, Brooks KA, Cohen S, Tenenholz B, Desai D *et al.* (2009). The prevalence of BRCA1 mutations among young women with triple-negative breast cancer. *BMC Cancer*, **9**: 86.

13. Wong-Brown MW, Meldrum CJ, Carpenter JE, Clarke CL, Narod SA, Jakubowska A, Rudnicka H, Lubinski J, Scott RJ. (2015). Prevalence of BRCA1 and BRCA2 germline mutations in patients with triple-negative breast cancer. *Breast Cancer Res Treat*, **150**(1): 71–80.

14. Domagala P, Hybiak J, Cybulski C, Lubinski J. (2017). BRCA1/2-negative hereditary triple-negative breast cancers exhibit BRCAness. *Int J Cancer*, **140**(7): 1545–1550.

15. Engel C, Rhiem K, Hahnen E, Loibl S, Weber KE, Seiler S, Zachariae S, Hauke J, Wappenschmidt B, Waha A *et al.* (2018). Prevalence of pathogenic BRCA1/2 germline mutations among 802 women with unilateral triple-negative breast cancer without family cancer history. *BMC Cancer*, **18**(1): 265.

16. Arai M, Yokoyama S, Watanabe C, Yoshida R, Kita M, Okawa M, Sakurai A, Sekine M, Yotsumoto J, Nomura H *et al.* (2018). Genetic and clinical characteristics in Japanese hereditary breast and ovarian cancer: First report after establishment of HBOC registration system in Japan. *J Hum Genet*, **63**(4): 447–457.

17. Li YT, Ni D, Yang L, Zhao Q, Ou JH. (2014). The prevalence of BRCA1/2 mutations of triple-negative breast cancer patients in Xinjiang multiple ethnic region of China. *Eur J Med Res*, **19**: 35.

18. Kotoula V, Fostira F, Papadopoulou K, Apostolou P, Tsolaki E, Lazaridis G, Manoussou K, Zagouri F, Pectasides D, Vlachos I *et al.* (2017). The fate of BRCA1-related germline mutations in triple-negative breast tumors. *Am J Cancer Res*, **7**(1): 98–114.

19. van Verschuer VM, Hooning MJ, van Baare-Georgieva RD, Hollestelle A, Timmermans AM, Koppert LB, Verhoog LC, Martens JW, Seynaeve C, van Deurzen CH. (2015). Tumor-associated inflammation as a potential prognostic tool in BRCA1/2-associated breast cancer. *Hum Pathol*, **46**(2): 182–190.

20. Bayraktar S, Gutierrez-Barrera AM, Liu D, Tasbas T, Akar U, Litton JK, Lin E, Albarracin CT, Meric-Bernstam F, Gonzalez-Angulo AM *et al.* (2011). Outcome of triple-negative breast cancer in patients with or without deleterious BRCA mutations. *Breast Cancer Res Treat*, **130**(1): 145–153.

21. Afghahi A, Timms KM, Vinayak S, Jensen KC, Kurian AW, Carlson RW, Chang PJ, Schackmann E, Hartman AR, Ford JM *et al.* (2017). Tumor BRCA1 reversion mutation arising during neoadjuvant platinum-based chemotherapy in triple-negative breast cancer is associated with therapy resistance. *Clin Cancer Res*, **23**(13): 3365–3370.

22. Bayraktar S, Gluck S. (2012). Systemic therapy options in BRCA mutation-associated breast cancer. *Breast Cancer Res Treat*, **135**(2): 355–366.

23. Garraway LA, Lander ES. (2013). Lessons from the cancer genome. *Cell*, **153**(1): 17–37.

24. Zhao S, Zuo WJ, Shao ZM, Jiang YZ. (2020). Molecular subtypes and precision treatment of triple-negative breast cancer. *Ann Transl Med*, **8**(7): 499.

25. Rechsteiner M, Dedes K, Fink D, Pestalozzi B, Sobottka B, Moch H, Wild P, Varga Z. (2018). Somatic BRCA1 mutations in clinically sporadic breast cancer with medullary histological features. *J Cancer Res Clin Oncol*, **144**(5): 865–874.

26. Pop LA, Cojocneanu-Petric RM, Pileczki V, Morar-Bolba G, Irimie A, Lazar V, Lombardo C, Paradiso A, Berindan-Neagoe I. (2018). Genetic alterations in sporadic triple negative breast cancer. *Breast*, **38**: 30–38.

27. Ma D, Jiang YZ, Liu XY, Liu YR, Shao ZM. (2017). Clinical and molecular relevance of mutant-allele tumor heterogeneity in breast cancer. *Breast Cancer Res Treat*, **162**(1): 39–48.

28. Dillon JL, Mockus SM, Ananda G, Spotlow V, Wells WA, Tsongalis GJ, Marotti JD. (2016). Somatic gene mutation analysis of triple negative breast cancers. *Breast*, **29**: 202–207.

29. Santarpia L, Bottai G, Kelly CM, Gyorffy B, Szekely B, Pusztai L. (2016). Deciphering and targeting oncogenic mutations and pathways in breast cancer. *Oncologist*, **21**(9): 1063–1078.

30. Guerini-Rocco E, Piscuoglio S, Ng CK, Geyer FC, De Filippo MR, Eberle CA, Akram M, Fusco N, Ichihara S, Sakr RA *et al.* (2016). Microglandular adenosis associated with triple-negative breast cancer is a neoplastic lesion of triple-negative phenotype harbouring TP53 somatic mutations. *J Pathol*, **238**(5): 677–688.

31. Ahmad F, Badwe A, Verma G, Bhatia S, Das BR. (2016). Molecular evaluation of PIK3CA gene mutation in breast cancer: Determination of

frequency, distribution pattern and its association with clinicopathological findings in Indian patients. *Med Oncol*, **33**(7): 74.

32. Weisman PS, Ng CK, Brogi E, Eisenberg RE, Won HH, Piscuoglio S, De Filippo MR, Ioris R, Akram M, Norton L *et al.* (2016). Genetic alterations of triple negative breast cancer by targeted next-generation sequencing and correlation with tumor morphology. *Mod Pathol*, **29**(5): 476–488.

33. Yamashita N, Tokunaga E, Kitao H, Hitchins M, Inoue Y, Tanaka K, Hisamatsu Y, Taketani K, Akiyoshi S, Okada S *et al.* (2015). Epigenetic inactivation of BRCA1 through promoter hypermethylation and its clinical importance in triple-negative breast cancer. *Clin Breast Cancer*, **15**(6): 498–504.

34. Knight JF, Lesurf R, Zhao H, Pinnaduwage D, Davis RR, Saleh SM, Zuo D, Naujokas MA, Chughtai N, Herschkowitz JI *et al.* (2013). Met synergizes with p53 loss to induce mammary tumors that possess features of claudin-low breast cancer. *Proc Natl Acad Sci USA*, **110**(14): E1301–1310.

35. Wen YH, Ho A, Patil S, Akram M, Catalano J, Eaton A, Norton L, Benezra R, Brogi E. (2012). Id4 protein is highly expressed in triple-negative breast carcinomas: Possible implications for BRCA1 downregulation. *Breast Cancer Res Treat*, **135**(1): 93–102.

36. Shah SP, Roth A, Goya R, Oloumi A, Ha G, Zhao Y, Turashvili G, Ding J, Tse K, Haffari G *et al.* (2012). The clonal and mutational evolution spectrum of primary triple-negative breast cancers. *Nature*, **486**(7403): 395–399.

37. Sanchez-Munoz A, Gallego E, de Luque V, Perez-Rivas LG, Vicioso L, Ribelles N, Lozano J, Alba E. (2010). Lack of evidence for KRAS oncogenic mutations in triple-negative breast cancer. *BMC Cancer*, **10**: 136.

38. Knight JF, Sung VYC, Kuzmin E, Couzens AL, de Verteuil DA, Ratcliffe CDH, Coelho PP, Johnson RM, Samavarchi-Tehrani P, Gruosso T *et al.* (2018). KIBRA (WWC1) is a metastasis suppressor gene affected by chromosome 5q loss in triple-negative breast cancer. *Cell Rep*, **22**(12): 3191–3205.

39. Assadipour Y, Zacharakis N, Crystal JS, Prickett TD, Gartner JJ, Somerville RPT, Xu H, Black MA, Jia L, Chinnasamy H *et al.* (2017). Characterization of an immunogenic mutation in a patient with metastatic triple-negative breast cancer. *Clin Cancer Res*, **23**(15): 4347–4353.

40. Lee JY, Park K, Lim SH, Kim HS, Yoo KH, Jung KS, Song HN, Hong M, Do IG, Ahn T *et al.* (2015). Mutational profiling of brain metastasis

from breast cancer: Matched pair analysis of targeted sequencing between brain metastasis and primary breast cancer. *Oncotarget*, 6(41): 43731–43742.

41. Xie Y, Gou Q, Wang Q, Zhong X, Zheng H. (2017). The role of BRCA status on prognosis in patients with triple-negative breast cancer. *Oncotarget*, 8(50): 87151–87162.

42. Brianese RC, Nakamura KDM, Almeida F, Ramalho RF, Barros BDF, Ferreira ENE, Formiga M, de Andrade VP, de Lima VCC, Carraro DM. (2018). BRCA1 deficiency is a recurrent event in early-onset triple-negative breast cancer: A comprehensive analysis of germline mutations and somatic promoter methylation. *Breast Cancer Res Treat*, 167(3): 803–814.

43. Avery-Kiejda KA, Morten B, Wong-Brown MW, Mathe A, Scott RJ. (2014). The relative mRNA expression of p53 isoforms in breast cancer is associated with clinical features and outcome. *Carcinogenesis*, 35(3): 586–596.

44. Zhong X, Dong Z, Dong H, Li J, Peng Z, Deng L, Zhu X, Sun Y, Lu X, Shen F *et al.* (2016). Prevalence and prognostic role of BRCA1/2 variants in unselected chinese breast cancer patients. *PloS One*, 11(6): e0156789.

45. Gonzalez-Angulo AM, Timms KM, Liu S, Chen H, Litton JK, Potter J, Lanchbury JS, Stemke-Hale K, Hennessy BT, Arun BK *et al.* (2011). Incidence and outcome of BRCA mutations in unselected patients with triple receptor-negative breast cancer. *Clin Cancer Res*, 17(5): 1082–1089.

46. Clement E, Inuzuka H, Nihira NT, Wei W, Toker A. (2018). Skp2-dependent reactivation of AKT drives resistance to PI3K inhibitors. *Sci Signal*, 11(521): pii: eaao3810.

47. Nolan E, Savas P, Policheni AN, Darcy PK, Vaillant F, Mintoff CP, Dushyanthen S, Mansour M, Pang JB, Fox SB *et al.* (2017). Combined immune checkpoint blockade as a therapeutic strategy for BRCA1-mutated breast cancer. *Sci Transl Med*, 9(393): pii: eaal4922.

48. Liu JC, Voisin V, Wang S, Wang DY, Jones RA, Datti A, Uehling D, Al-awar R, Egan SE, Bader GD *et al.* (2014). Combined deletion of Pten and p53 in mammary epithelium accelerates triple-negative breast cancer with dependency on eEF2K. *EMBO Mol Med*, 6(12): 1542–1560.

49. Cossu-Rocca P, Orru S, Muroni MR, Sanges F, Sotgiu G, Ena S, Pira G, Murgia L, Manca A, Uras MG *et al.* (2015). Analysis of PIK3CA muta-

tions and activation pathways in triple negative breast cancer. *PloS One*, **10**(11): e0141763.

50. Katoh M. (2007). WNT signaling pathway and stem cell signaling network. *Clin Cancer Res*, **13**(14): 4042–4045.

51. Zhang MZ, Ferrigno O, Wang Z, Ohnishi M, Prunier C, Levy L, Razzaque M, Horne WC, Romero D, Tzivion G *et al.* (2015). TGIF governs a feed-forward network that empowers Wnt signaling to drive mammary tumorigenesis. *Cancer Cell*, **27**(4): 547–560.

52. Monteiro J, Gaspar C, Richer W, Franken PF, Sacchetti A, Joosten R, Idali A, Brandao J, Decraene C, Fodde R. (2014). Cancer stemness in Wnt-driven mammary tumorigenesis. *Carcinogenesis*, **35**(1): 2–13.

53. Trifa F, Karray-Chouayekh S, Jmal E, Jmaa ZB, Khabir A, Sellami-Boudawara T, Frikha M, Daoud J, Mokdad-Gargouri R. (2013). Loss of WIF-1 and Wnt5a expression is related to aggressiveness of sporadic breast cancer in Tunisian patients. *Tumour Biol*, **34**(3): 1625–1633.

54. Ma X, Zhao X, Yan W, Yang J, Zhao X, Zhang H, Hui Y, Zhang S. (2018). Tumor-infiltrating lymphocytes are associated with beta-catenin overexpression in breast cancer. *Cancer Biomark*, **21**(3): 639–650.

55. Daly CS, Shaw P, Ordonez LD, Williams GT, Quist J, Grigoriadis A, Van Es JH, Clevers H, Clarke AR, Reed KR. (2017). Functional redundancy between Apc and Apc2 regulates tissue homeostasis and prevents tumorigenesis in murine mammary epithelium. *Oncogene*, **36**(13): 1793–1803.

56. Yu X, Wang M, Dong Q, Jin F. (2014). Diversin is overexpressed in breast cancer and accelerates cell proliferation and invasion. *PloS One*, **9**(5): e98591.

57. Bilir B, Kucuk O, Moreno CS. (2013). Wnt signaling blockage inhibits cell proliferation and migration, and induces apoptosis in triple-negative breast cancer cells. *J Transl Med*, **11**: 280.

58. Wend P, Runke S, Wend K, Anchondo B, Yesayan M, Jardon M, Hardie N, Loddenkemper C, Ulasov I, Lesniak MS *et al.* (2013). WNT10B/ beta-catenin signalling induces HMGA2 and proliferation in metastatic triple-negative breast cancer. *EMBO Mol Med*, **5**(2): 264–279.

59. Hou MF, Chen PM, Chu PY. (2018). LGR5 overexpression confers poor relapse-free survival in breast cancer patients. *BMC Cancer*, **18**(1): 219.

60. Mukherjee N, Dasgupta H, Bhattacharya R, Pal D, Roy R, Islam S, Alam N, Biswas J, Roy A, Roychoudhury S *et al.* (2016). Frequent inactivation of MCC/CTNNBIP1 and overexpression of phospho-beta-catenin(Y654) are

associated with breast carcinoma: Clinical and prognostic significance. *Biochim Biophys Acta*, **1862**(9): 1472–1484.

61. Lakis S, Dimoudis S, Kotoula V, Alexopoulou Z, Kostopoulos I, Koletsa T, Bobos M, Timotheadou E, Papaspirou I, Efstratiou I *et al.* (2016). Interaction between beta-catenin and egfr expression by immunohisto-chemistry identifies prognostic subgroups in early high-risk triple-negative breast cancer. *Anticancer Res*, **36**(5): 2365–2378.

62. Yang L, Perez AA, Fujie S, Warden C, Li J, Wang Y, Yung B, Chen YR, Liu X, Zhang H *et al.* (2014). Wnt modulates MCL1 to control cell survival in triple negative breast cancer. *BMC Cancer*, **14**: 124.

63. Xu WH, Liu ZB, Yang C, Qin W, Shao ZM. (2012). Expression of dick-kopf-1 and beta-catenin related to the prognosis of breast cancer patients with triple negative phenotype. *PloS One*, **7**(5): e37624.

64. Lu H, Tran L, Park Y, Chen I, Lan J, Xie Y, Semenza GL. (2018). Reciprocal regulation of DUSP9 and DUSP16 expression by HIF1 controls ERK and p38 MAP kinase activity and mediates chemotherapy-induced breast cancer stem cell enrichment. *Cancer Res*, **78**(15): 4191–4202.

65. Majorini MT, Manenti G, Mano M, De Cecco L, Conti A, Pinciroli P, Fontanella E, Tagliabue E, Chiodoni C, Colombo MP *et al.* (2018). cIAP1 regulates the EGFR/Snai2 axis in triple-negative breast cancer cells. *Cell Death Differ*, **25**(12): 2147–2164.

66. Chen Y, Wang X, Cao C, Wang X, Liang S, Peng C, Fu L, He G. (2017). Inhibition of HSP90 sensitizes a novel Raf/ERK dual inhibitor CY-9d in triple-negative breast cancer cells. *Oncotarget*, **8**(61): 104193–104205.

67. Nagaria TS, Shi C, Leduc C, Hoskin V, Sikdar S, Sangrar W, Greer PA. (2017). Combined targeting of Raf and Mek synergistically inhibits tumorigenesis in triple negative breast cancer model systems. *Oncotarget*, **8**(46): 80804–80819.

68. Dushyanthen S, Teo ZL, Caramia F, Savas P, Mintoff CP, Virassamy B, Henderson MA, Luen SJ, Mansour M, Kershaw MH *et al.* (2017). Agonist immunotherapy restores T cell function following MEK inhibi-tion improving efficacy in breast cancer. *Nat Commun*, **8**(1): 606.

69. Van Swearingen AED, Sambade MJ, Siegel MB, Sud S, McNeill RS, Bevill SM, Chen X, Bash RE, Mounsey L, Golitz BT *et al.* (2017). Combined kinase inhibitors of MEK1/2 and either PI3K or PDGFR are efficacious in intracranial triple-negative breast cancer. *Neuro Oncol*, **19**(11): 1481–1493.

70. Liang Y, Zhou Y, Deng S, Chen T. (2016). Microwave-assisted syntheses of benzimidazole-containing selenadiazole derivatives that induce cell-cycle arrest and apoptosis in human breast cancer cells by activation of the ROS/AKT pathway. *ChemMedChem*, 11(20): 2339–2346.

71. Noman AS, Uddin M, Rahman MZ, Nayeem MJ, Alam SS, Khatun Z, Wahiduzzaman M, Sultana A, Rahman ML, Ali MY *et al.* (2019). Overexpression of sonic hedgehog in the triple negative breast cancer: Clinicopathological characteristics of high burden breast cancer patients from Bangladesh. *Sci Rep*, 6: 18830.

72. Arnold KM, Pohlig RT, Sims-Mourtada J. (2017). Co-activation of hedgehog and Wnt signaling pathways is associated with poor outcomes in triple negative breast cancer. *Oncol Lett*, 14(5): 5285–5292.

73. Riaz SK, Khan JS, Shah STA, Wang F, Ye L, Jiang WG, Malik MFA. (2018). Involvement of hedgehog pathway in early onset, aggressive molecular subtypes and metastatic potential of breast cancer. *Cell Commun Signal*, 16(1): 3.

Chapter FOUR

Epigenetics in Triple-Negative Breast Cancer

Ju Zhou[1], Sisi Chen[2,3], Md. Asaduzzaman Khan[1], Ting Xiao[1], Mousumi Tania[4], Md. Shamsuddin Sultan Khan[5], *and* Junjiang Fu[1,*]

Contents

*Corresponding author: Junjiang Fu, E-mail: fujunjiang@hotmail.com

[1]Key Laboratory of Epigenetics and Oncology, Research Center for Preclinical Medicine, Southwest Medical University, Luzhou, Sichuan, China.

[2]Key Laboratory of Translational Cancer Stem Cell Research, Hunan Normal University, Changsha, Hunan, China.

[3]Departments of Pathology and Pathophysiology, Hunan Normal University School of Medicine, Changsha, Hunan, China.

[4]Division of Molecular Cancer Biology, The Red-Green Research Center, Dhaka, Bangladesh.

[5]EMAN Testing & Research Laboratory, Department of Pharmacology, School of Pharmaceutical Sciences, Universiti Sains Malaysia, Minden, Penang, Malaysia.

Epigenetics is the study of heritable changes in phenotypes that do not involve any change in DNA sequences. In 1942, Conrad H. Waddington coined the words "epigenesis" and "genetics" to describe the "causal mechanisms" by which "the genes of the genotype bring about phenotypic effects" [1]. Due to the lack of experimental tools and overall knowledge, it took over 50 years for scientists to understand the underlying mechanisms of Waddington's observations [2]. To date, multiple discoveries have been made describing how epigenetics can change a phenotype without altering DNA sequences. Epigenetics affects the entire genome but, with regards to cancer, it tends to affect key oncogenes, tumor suppressor genes, and transcription factors, leading to cancer initiation and progression [3]. Epigenetics is brought about via different heritable and reversible mechanisms such as DNA methylation, histone modification, and chromatin remodeling, as well as the more recently discovered epigenetic changes through non-coding RNAs (discussed later) [4]. Among these epigenetic mechanisms, the most common modifications are DNA methylation, histone lysine methylation and histone lysine acetylation [5].

A large amount of data showed distinct epigenomic profiles that distinguish breast cancers from normal and benign tissues. Hence, taking advantage of the reversibility of epigenetic modifications, drugs that target epigenetic modifiers, given in combination with chemotherapies or endocrine therapies, may present promising approaches to restoration of therapy responsiveness in cancer [3]. There remain many open questions about the mechanisms involved in epigenetic control, but it is recognized that epigenetic change can occur due to environmental factors such as stress and cell damage. However, little is known about why and how genes are activated only when they are required. Of note, an increase in the understanding of epigenetic mechanisms and their contributions to disease development has led to a growing interest in the field of epigenetics [6]. Recently, scientists have demonstrated a significant impact of epigenetic alterations on triple-negative breast cancer (TNBC).

4.1 Alterations in DNA Methylation in TNBC

4.1.1 *Basics of DNA Methylation*

The first recognized and most well-characterized epigenetic modification in mammals is DNA methylation which occurs in the cytosine residue in CpG dinucleotides. DNA methylation is a process by which methyl groups are added to the DNA molecule (**Figure 4-1**). The covalent addition of a methyl group to the cytosine, which results in

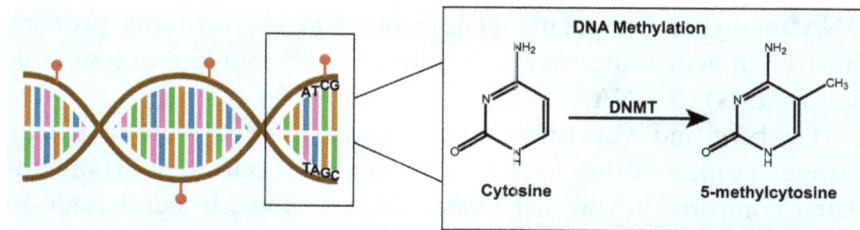

Figure 4-1. The process of DNA methylation. Through the action of DNA methyltransferase, a methyl group at the carbon-5 position of the cytosine residue within the CpG dinucleotide is added. DNMT, DNA methyltransferase.

the formation of 5'-methylcytosine (5mC) in a CpG dinucleotide, is catalyzed by a family of enzymes named DNA methyltransferases (DNMTs). DNMTs catalyze the addition of a methyl group to the carbon-5 position of cytosine bases into 5mC, which generally leads to gene silencing [6]. While DNMT1 is involved in the maintenance of DNA methylation after replication, DNMT3A and DNMT3B are mostly involved in the de novo methylation of DNA. Cytosine methylation is widespread in both eukaryotes and prokaryotes [7].

DNA methylation mostly occurs on CpG islands, which are DNA regions with at least 200 bases that consist of at least 50% C+G content [8]. The majority of human promoters contain CpG islands and are usually hypomethylated. Only a few of these CpG islands become methylated during development or cell differentiation. Different DNMTs are also necessary to maintain DNA methylation after cells complete a cell division. DNMT1 will copy methylation patterns and replicate it to the daughter DNA strand and is therefore regarded as a maintenance enzyme [9]. In mammals, the enzymes DNMT3a and DNMT3b are responsible for the initial DNA methylation [10]. Importantly, DNA methylation can inhibit gene expression in various ways. It can lead to binding of methyl-CpG-binding domain (MBD) proteins, which then recruit histone modifying and/or chromatin remodeling complexes to the methylated site to compact and inactivate chromatin [11]. DNA methylation can also inhibit the binding of transcription factors to promoters, which, nevertheless, does not occur frequently [12, 13].

Importantly, methylation can change the activity of a DNA segment without changing its sequence [7]. In general, hypermethylation in the promoter of genes causes their transcriptional silencing. DNA methylation regulates gene expression by recruiting proteins involved in gene repression or by inhibiting the binding of transcription factor(s) to DNA.

Feinberg and Vogelstein were the first to report on epigenetic change in cancer as they found colorectal cancer cells were hypomethylated compared to normal tissue. DNA hypomethylation leads to oncogene activation and chromosome instability, culminating in tumor development. Conversely, hypermethylation has been shown to inhibit tumor suppressor genes, thereby releasing cells from their normal control. DNA methylation is usually associated with genomic stability, but is also associated with the control of gene expression via an alteration in the accessibility of transcriptional start sites. Hypermethylation of tumor suppressor genes, including p16, death-associated protein kinase (DAPK), E-cadherin, and O^6-methylguanyl DNA methyltransferase (MGMT) is associated with poorer prognosis in various cancer types. In glioblastoma, the methylated promoter of MGMT is a predictive marker of response to alkylating agents such as temozolomide, as well as a predictive marker of radiotherapy response in the absence of alkylating chemotherapy. Importantly, overexpression of DNMT genes has been reported in various cancers [14]. Promoter hypermethylation of a number of tumor-associated genes, detected in body fluids of cancer patients, has emerged as a promising biomarker for the early detection of cancers, including breast cancer [5]. Although epigenetic profiling is not yet commonly performed, the FDA approved the first screening test, Cologuard, based on the analysis of DNA methylation for colorectal cancer in August 2014 [15].

4.1.2 *DNA Methylation Signatures in TNBC*

In Chapter 2, it has been discussed that current efforts to stratify early breast cancer prognosis have primarily focused on multi-gene expression signatures. Such signatures are most effective at assigning recurrence risk to early-stage hormone receptor-positive breast cancer, in which the rate of proliferation is closely associated with overall prognosis. However, the majority of TNBCs are highly proliferative and therefore cannot be stratified using these multi-gene classifiers [16]. In addition to multi-gene expression assays, DNA methylation signatures are being assessed as potential molecular biomarkers of cancer. A number of studies have documented aberrant methylation events in breast carcinogenesis and the specifically identified DNA methylation biomarkers that have significant diagnostic and prognostic potential. Several studies have also identified DNA methylation signatures that can distinguish between breast cancer subtypes and others that may be predictive of treatment response [7].

Analysis of cancer methylomes has dramatically changed the concept of the potential of diagnostic and prognostic methylation biomarkers in disease stratification [16]. Differentially methylated regions (DMRs) are genomic regions with different DNA methylation status across different biological samples and regarded as possible functional regions involved in gene transcriptional regulation. Through whole-genome methylation capture-sequencing of TNBCs, scientists recently identified DMRs with diagnostic and prognostic values that stratify TNBCs for more personalized management.

By studying the DNA methylation group of TNBC, scientists identified 865 DMRs in TNBC tumors compared to matched normal samples and showed that these regions were enriched in promoters associated with transcription factor binding sites and DNA hypersensitive sites. Strikingly, in this study, researchers found 36 DMRs that were specific to TNBC and also showed that these DMRs stratified TNBC patients into 3 distinct methylation clusters. Survival analysis revealed that the largely hypomethylated cluster was associated with better prognosis, whereas the other 2 more methylated clusters were associated with worse prognosis. Additional survival analysis identified 17 DMRs in the cancer genome atlas (TCGA) breast cancer data, each harboring 3 or more HM450K probes associated with survival in TNBC samples; 14 genomic regions were associated with poor survival and 3 regions with longer survival. Kaplan-Meier plots of individual CpG sites in each region showed very good survival separation, highlighting their potential value as prognostic biomarkers. Interestingly, most of these regions overlap with DNase I hypersensitive sites and contain many transcription factor binding sites, suggesting that they may harbor important regulatory functions [17]. This study has therefore provided the first piece of evidence that DNA methylation profiling can be used to classify breast cancer subtypes and stratify TNBCs according to patient outcome.

DNMT inhibitors (DNMTi's) are able to reverse DNA hypermethylation through removing methyl groups from the promoter of the target gene, thus reactivating the "silenced" gene. Currently, 5-azacytidine, a chemical analog of cytidine, and its deoxy derivative, decitabine (also known as 5-aza-2'-deoxycytidine), are two commonly used DNMTi's. These DNMTi's have been approved for the treatment of myelodysplastic syndrome (MDS) and acute myeloid leukemia (AML). The potential role of DNMTi's in TNBC treatment will be discussed in more detail in Chapter 7.

4.2 Alterations in Histone Modifications in TNBC

4.2.1 *Reversible Histone Acetylation/Deacetylation*

In eukaryotic cells, the organization of DNA into chromatin is necessary for the preservation of genomic integrity and is required for the correct transmission of genetic information over generations. In addition to the physical role of compacting and protecting DNA, the chromatin conformation is closely correlated with the expression state of the genes within its structure. Genes present in a dense chromatin environment are less available to the transcriptional machinery and transcribed to a lesser extent than genes found in looser and more permissive, chromatin domains. As stated above, chromatin is subject to highly dynamic modifications, playing important roles in regulating the availability of DNA and thus gene expression. This regulation includes the exchange of histone variants, nucleosome remodeling by ATP-dependent remodeling complexes, as well as post-translational modifications of DNA and histones. The basic structural unit of chromatin is nucleosome. The nucleosome contains a DNA superhelix of about 200 bp and a histone octamer, which ensures that DNA is highly concentrated. Post-translational histone modifications (PTHMs) will change this structure, thus affecting gene transcription. Classically, PTHMs include acetylation, methylation, ubiquitination, phosphorylation, and sulfonation. In recent times, several other novel PTHMs have been observed, such as neddylation, glycosylation, poly-ADP ribosylation, as well as transcriptional regulation [18]. Among them, the most characteristic is acetylation and it seems more closely related to the development of TNBC.

Histones have a highly dynamic role in the regulation of chromatin structure and gene activity. Histone tails can be modified by acetylation, methylation, phosphorylation, poly-ADP ribosylation, sumoylation, ubiquitination, etc. A combination of histone modifications and histone variants, referred to as the "histone code", determines the interaction of histones with DNA and the interaction of non-histone proteins with chromatin. Histone acetylation by histone acetyltransferases (HATs) neutralizes the positive charge on lysine residues in the histones, thereby loosening their interaction with DNA, thus rendering the chromatin more accessible to transcription factors. Several HATs were found to have an important role in a variety of cancers [19]. Histone deacetylases (HDACs) remove lysine acetyl groups, thus compacting the conformation of chromatin (**Figure 4-2**). Aberrant expressions of HDACs were found in a variety of cancers, including breast cancer.

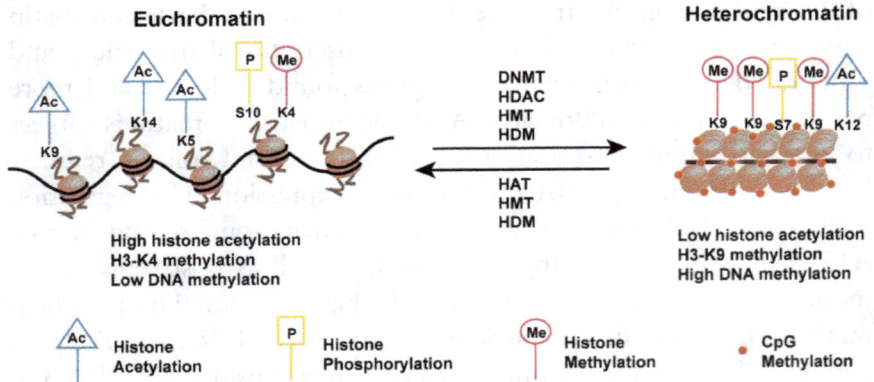

Figure 4-2. Mechanisms of epigenetic transcriptional regulation involving DNA methylation and histone modifications. In euchromatin, CpG islands in the gene promoters are unmethylated, thus allowing for HATs to be recruited, which promotes transcriptional activation through elevated histone acetylation, while H3-K4 methylation blocks DNMTs binding. Condensed heterochromatin contains CpG methylation, catalyzed by methylation "writers" DNMTs; methyl-binding proteins (MBPs) bind to the methylated DNA, act as methylation "readers" that recruit transcription repression complexes, including HDACs and other histone modifying enzymes, and thus gene expression is silenced. DNMT, DNA methyltransferase; HDAC, histone deacetylase; HMT, histone methyltransferase; HDM, histone demethylase; HAT, histone acetyltransferase.

Acetylation/deacetylation of histones is regulated by the balanced activity of HATs and HDACs. As stated earlier, deacetylated histones are more negatively charged, thus binding DNA tightly and this condensed heterochromatin is less accessible to transcriptional factors and other regulatory transcription machinery proteins. Both histone acetylation "writers" — HATs and histone acetylation "erasers" — HDACs are required for epigenetic regulation of gene expression [14]. The aberrant global and gene-specific histone acetylation patterns and deregulated expression of HATs and HDACs are implicated in malignant transformation. In many cancer types, promoter hypoacetylation has been associated with repressed gene expression of tumor suppressor genes.

In humans, eighteen HDACs have been identified and are generally divided into four classes based on sequence homology to yeast counterparts, intracellular localization and enzymatic activity: Class I, Class II (divided into Class IIa and Class IIb: HDAC6,10), Class III (sirtuins (SIRTs) 1–7) and Class IV (HDAC11) [14]. Although some HDACs exert similar biological effects, they may also have highly specific roles in specific cancer types. For example, HDAC4 expression is upregulated in breast cancer samples compared to renal, bladder and colorectal cancer, which indicates differential expression of the selected HDACs in human solid cancers. These findings suggested that transcriptional repression of tumor-suppressor genes by overexpression and aberrant recruitment of HDACs to their promoter regions could be a common phenomenon in tumor onset and progression. Thus, HDACi's emerge as valuable and attractive therapeutic agents in cancer treatment.

4.2.2 *Histone Deacetylase Inhibitors and Their Potential Applications in TNBC*

Currently used HDAC inhibitors (HDACi's) include vorinostat and romidepsin. Interestingly, plant-derived bioactive compounds with anti-cancer properties, which show inhibitory effects on DNMTs as mentioned above, have also been found to inhibit

HDACs. Due to the promising results in preclinical studies, vorinostat and romidepsin are being tested in several clinical trials for breast cancer and other solid tumors. Some studies have sought to provide detailed information that describes the epigenomic and transcriptomic effects of HDACi treatment. These effects include genome-wide chromatin accessibility, DNA methylation, and gene expression alterations after HDACi treatment of two human TNBC cell lines [20].

A recent study investigated the synergistic effect of a novel HDACi, OBP-801, and eribulin (a microtubule dynamics inhibitor used to treat metastatic breast cancer) in TNBC cell lines since OBP-801 is known to enhance the anti-tumor activities of other chemotherapeutic agents. This study showed that the combination treatment with OBP-801 and eribulin synergistically inhibited growth and increased TNBC cell apoptosis. Moreover, this was the first indication that eribulin upregulates the anti-apoptotic protein Survivin, which could be remarkably suppressed by OBP-801. Also, this combination potently suppressed Bcl-xL and the MAPK pathway compared with either agent alone [21].

The potential strategy that targets epigenetics in TNBC will be discussed in more detail in Chapter 7. It should be noted that the best clinical efficacy for epigenetic therapy has been achieved in patients with hematologic malignancies. In current epigenetic therapy, however, treatment with DNMTi's or HDACi's as single agents has limited clinical benefit for patients with solid tumors. Combination of DNMTi's with HDACi's has been evaluated in cancer therapy and has been shown to be effective in overcoming tamoxifen resistance in breast cancer [22]. Another trend is the use of epigenetic drugs in combination with chemotherapeutic drugs or radiotherapy. This concept has been extended by combining epigenetic therapy with immunotherapy. It has been demonstrated that azacytidine that alters the epigenome primes the immune system of the patient to respond to the immune checkpoint inhibitor, e.g., nivolumab (a monoclonal antibody targeting cell surface protein programmed cell death-1 (PD1)). Immunotherapy of TNBC based on immune checkpoint inhibition will be discussed in Chapter 8.

4.3 Alterations in Phosphatidylcholine Metabolism in TNBC

4.3.1 *Reprogramming of Phosphatidylcholine Metabolism in TNBC*

Phospholipids play a dual role of being basic structural components of the plasma membrane and acting as substrates of reactions involved in key regulatory functions in mammalian cells [23, 24]. Phosphatidylcholine (PtdCho), the most abundant phospholipid in eukaryotic cell membrane, is synthesized via the Kennedy pathway. The initial step in the Kennedy pathway is uptake of choline (Cho) into cells through the facilitative transporters, i.e., choline transporter-like protein (CTL), organic cation transporter-2 (OCT2), and choline high-affinity transporter-1 (CHT1). An ATP-dependent choline kinase (ChoK) synthesizes phosphocholine (PCho), which is then converted to cytidine diphosphate choline (CDPCho) by cytidylyltransferase (CCT) [25]. Choline phosphotransferase (CPT) catalyzes the synthesis of PtdCho by transfer of the PCho moiety from CDPCho to the 3-hydroxyl group of diacylglycerol (DAG) [25]. Hydrolysis of PtdCho can generate second messengers, such as DAG, phosphatidic acid (PA), and lysophosphatidylcholine (LPtdCho). These PtdCho metabolites are produced through three major catabolic pathways, respectively, mediated by phospholipase C (PLC) and D (PLD), which act at the two distinct phosphodiester bonds of the PtdCho headgroup and by phospholipase A2 (PLA2) and A1 (PLA1), which act in the deacylation reaction cascade (**Figure 4-3**) [26–31].

The concerted activation of assembly of molecular complexes in cancer cells cooperates to sustain oncogene-induced cell signaling through multiple intracellular pathways involved in phospholipid biosynthesis and breakdown. Among these, phosphatidylinositol 4-phosphate 5-kinase Iγ (PIPKIγ) is overexpressed in TNBC cells [32]. Furthermore, two major enzymes involved in the agonist-induced PtdCho cycle, such as ChoK and PtdCho-specific phospholipase C (PC-PLC), are overexpressed and activated in various breast cancer subtypes, including TNBC [27, 33–36]. The present evidence

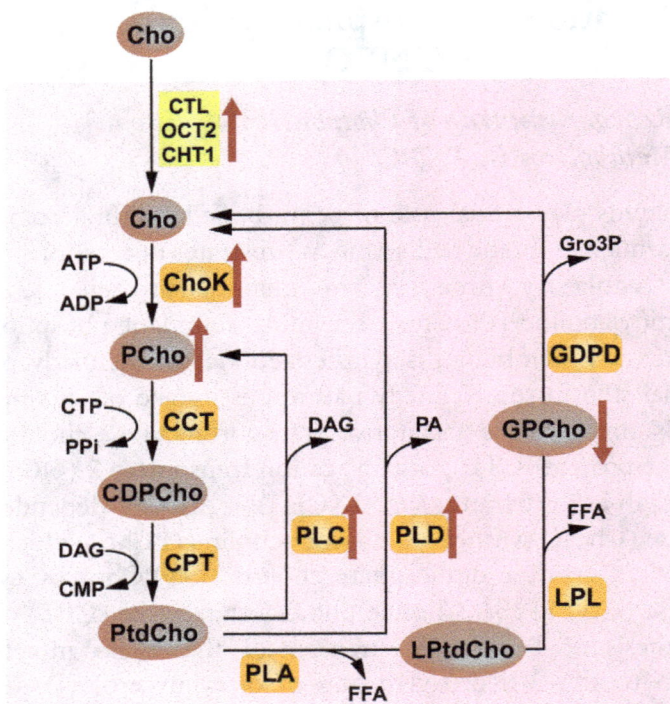

Figure 4-3. PtdCho cycle in TNBC. In TNBC cells, the activity or expression of Cho-metabolizing enzymes (ChoK, PLC, and PLD) and Cho transporters (CTL, OCT2, and CHT1) are upregulated and the content of PCho is increased, while the content of GPCho is reduced. Cho, free choline; PCho, phosphocholine; CDPCho, cytidine diphosphate choline; PtdCho, phosphatidylcholine; LPtdCho, lysophosphatidylcholine; GPCho, glycerophosphocholine; ChoK, choline kinase; CCT, phosphocholine cytidylyltransferase; CPT, choline phosphotransferase; PLC, phospholipase C; PLD, phospholipase D; PLA, phospholipase A; LPL, lysophospholipase; GDPD, glycerophosphocholine phosphodiesterase; CTL, choline transporter-like protein; OCT2, organic cation transporter-2; CHT1, choline high-affinity transporter-1; DAG, diacylglycerol; PA, phosphatidic acid; FFA, free fatty acid; Gro3P, sn-glycerol-3-phosphate. Red arrows indicate direction of change in metabolite and enzyme content, and enzyme activity.

points to the existence of multiple links between enzymes involved in the glycolytic gene/protein signature and those responsible for enhanced carbon fluxes through oncogene-driven PtdCho biosynthesis and catabolism in breast cancer cells. This biochemical interplay may also serve as a key regulator of tumor progression in TNBC [31].

Accumulation of phosphocholine, produced either by ChoK in the first reaction of the three-step Kennedy biosynthetic pathway or by PLC-mediated PtdCho catabolism, is associated with tumor growth and progression. This supports the inclusion of altered phospholipid metabolism as a novel candidate hallmark for cancer and as a key regulator in the overall cancer metabolic reprogramming program. Aberrant PtdCho metabolism associated with increases in the contents of the intracellular total choline (tCho) and PCho were initially observed in breast cancer cells as they progressed from normal to malignant phenotypes, i.e., from non-malignant immortalized cells to the highly metastatic cancer cells [37].

Several studies have shown altered PtdCho metabolism in TNBC, both in patients and in experimental models. Upregulation of ChoKα is a major contributor to the increased PCho content detected in TNBC. Phospholipase-mediated PtdCho headgroup hydrolysis also contributes to the buildup of a PCho pool in TNBC cells. The oncogene-driven PtdCho cycle appears to be finely tuned in TNBC cells in at least three ways: by modulating the Cho import, by regulating the activity or expression of specific metabolic enzymes and by contributing to the rewiring of the entire metabolic network.

Notably, a different rate of increase was observed for PCho (6-fold) and PtdCho (1.5-fold) in breast cancer cells compared with the non-malignant cells [38]. An integration of magnetic resonance spectroscopy (MRS) with gene microarray analysis revealed that a combination of upregulated ChoK and PLD and/or an increased PC-PLC expression/activity caused PCho accumulation in MDA-MB-231 cells, while lower levels of glycerophosphocholine were consistent with underexpression of cytosolic calcium-dependent PLA2 group IV A and lysophospholipase 1 [38].

A study conducted by Eliyahu *et al.* confirmed altered PtdCho metabolism in MDA-MB-231 TNBC cells compared with MCF-12A human mammary epithelial cells (HMECs) [39]. Interestingly, under the adopted experimental conditions, the PCho levels were found to correlate with Cho transport in the cells, mainly due to OCT2 and CHT1, but not with ChoK activity. The upregulation of Cho transporters and ChoK may be related to a cascade of genetic changes that are associated with the multistep process of carcinogenesis [40].

Enzymatic assays showed a 2- to 6-fold increase in the activation of PC-PLC in breast cancer of different subtypes compared with the non-tumoral counterpart [35]. The activity this enzyme measured in TNBC cells was about 2-fold higher than that of non-TNBC cells. Metabolic analysis of MDA-MB 231 cells identified a characteristic biochemical signature to be relative to the non-tumoral MCF-10A cells [41]. The importance of cell membrane lipid profiling to discriminate breast cancer subtypes is receiving increasing attention and the data suggest possible links between altered metabolic pathways in breast cancer and membrane molecular rearrangement [31].

4.3.2 *Efforts to Target Phosphatidylcholine Metabolism in TNBC*

The role of PtdCho cycle enzymes as potential new molecular targets in TNBC can be investigated using molecular depletion approaches and/or pharmacological inhibitors. Downregulation of ChoKα by RNA interference increased PLD expression, while its downregulation increased ChoKα expression, indicating a close relationship between ChoK and PLD enzymes [42]. Additionally, ChoKα silencing resulted in increased PC-PLC protein expression, suggesting that breast cancer cells could compensate for the loss of ChoKα protein levels with PC-PLC upregulation, thus maintaining an intracellular PCho pool size markedly higher than that of non-tumoral breast epithelial cells [31].

Interestingly, in TNBC cells, a combination of ChoK silencing with conventional treatment using 5-fluorouracil resulted in higher cell death rate than when each treatment was applied individually [43]. Additionally, there is mounting evidence from the studies on experimental TNBC models that the reduction/destabilization of ChoK protein levels rather than inhibition of the activity of this enzyme are more effective in inhibiting tumor growth. In fact, pharmacological inhibitors that were able to reduce the activity of ChoK did not reduce cell viability as long as ChoKα protein expression and PtdCho levels were not reduced in TNBC cells grown in vitro [44].

Although evidence of specific metabolic alterations in TNBC is accruing, there is a clear need for extending preclinical investigations to clinical TNBC models. Clinical investigations have to better elucidate the impact of the heterogeneous nature of TNBC lesions on the metabolic profiles and their changes in tumor progression. It may also prove relevant to assess the links between the tCho profile and molecular features such as EGFR overexpression, p53 status, and other biological characteristics of TNBC. The PtdCho cycle may present a good point of focus for personalized/precision medicine, offering markers that may be used as diagnosis tools for the assessment of cancer prognosis and response to therapy [31].

The identification of a role for PtdCho metabolism in TNBC progression supports the view that some enzymes of this cycle may act as key regulators of molecular mechanisms leading to cancer onset, invasion, and metastasis, thus representing a new source of potential targets to counteract cancer growth and progression [31].

4.4 Roles of Non-coding RNAs in TNBC

We have previously been under the notion that the transcriptome is mainly composed of the protein-coding sequences, i.e., mRNAs; however, high-throughput sequencing and the encyclopedia of DNA Elements (ENCODE) project have found that non-coding RNA is actually the principal constituent of the transcriptome. Besides the well-known RNAs such as tRNAs and rRNAs, many other regulatory RNAs have been identified. In humans and other mammals, protein-coding gene sequences represent only a minority (less than 2%) of the whole genome sequences; in contrast, the majority is represented by protein non-coding sequences, such as non-coding RNAs (ncRNAs). The ncRNAs can be divided into two categories: house-keeping ncRNAs (tRNA, rRNA, etc.) and regulatory ncRNAs (miRNA, lncRNA, circRNA, siRNA, piRNA, etc.).

Most ncRNAs are found to be effective in developing a highly complex RNA network important for gene transcription and translation, cell differentiation, hemogenesis, and heredity. NcRNAs regulate all these biological processes via RNA-DNA, RNA-RNA and

RNA-protein interactions. NcRNAs participate in almost all epigenetic regulations, including DNA methylation, imprinting, transposition, position effect variegation, chromatin modification, histone methylation and acetylation. Intense investigation on ncRNAs could provide novel tools for studying and treating human diseases such as cancer.

As important regulators of gene expression, ncRNAs have been found to play broad functions in different oncogenic and tumor suppressive pathways. NcRNAs can act as either oncogenes or tumor-suppressing regulators in key signaling pathways in tumor cells, thereby affecting tumorigenesis and metastasis. Consequently, ncRNAs have been proposed as novel biological tumor markers. NcRNAs play critical regulatory roles in tumor initiation, progression, and resistance to therapies. Understanding the roles of ncRNAs in cellular signaling network, particularly their clinical significance in tumorigenesis and progression, is the major challenge in cancer biology [45].

4.4.1 *MicroRNAs*

MicroRNAs (miRNAs or miRs) are small non-coding regulatory RNAs that contain roughly 21 to 25 nucleotides and play an essential role in cell signaling pathways [45–47]. MiRNAs are endogenously expressed small RNA sequences that act as post-transcriptional regulators of gene expression. They have been extensively studied for their roles in different cancers, especially for their ability to behave like oncogenes or tumor suppressors.

In 1993, the first miRNA lin-4 was discovered in Caenorhabditis elegans, where it was shown to decrease the levels of the Lin-4 protein by binding to the 3'-UTR region of its respective mRNA sequence [48]. Since that ground-breaking finding, miRNAs have been found to be highly conserved between species, suggesting that they play a universal role in the regulation of gene expression.

MiRNAs are generated endogenously through a series of steps, RNA polymerase II (or sometimes III) transcribes miRNAs in the nucleus as primary transcripts, pri-miRNA (~500–3,000 nucleotides). Drosha and DGCR8 shorten the pri-miRNA to ~70 nucleotides and build a stem-loop, which is called precursor miRNA (pre-miRNA). Exportin 5 transfers the pre-miRNA into the cytoplasm, where Dicer cuts it into 22-nucleotide RNA duplexes. In most cases, the strand with less paired bases on the 5′ end is the mature miRNA, whereas the other strand is degraded. The mature miRNA builds a complex with the Argonaute 2 protein and the heterodimer of R2D2 & Dicer-2 proteins to form the RNA-induced silencing complex (RISC) [49–52]. The RISC is able to silence the expression of a target gene, by binding to the 3'-UTR of the target gene (mRNA). The binding inhibits the ribosome from translating the gene, which leads to reduced expression of the target gene. Interested readers may refer to excellent reviews on this topic [51, 53].

More than 2,500 mature miRNAs have been identified in humans (miRBase v.21), but the functionality of most of them is yet to be discovered [54]. One miRNA can interact with multiple (>100) target genes and one gene can be controlled by multiple miRNAs [55]. More than 60% of all protein coding genes have conserved miRNA binding sites in their 3'-UTR region, which affords them the possibility of control by their respective miRNAs [56].

There are three possible ways that miRNAs can negatively affect the expression of its target mRNA (**Figure 4-4**). First, if the base

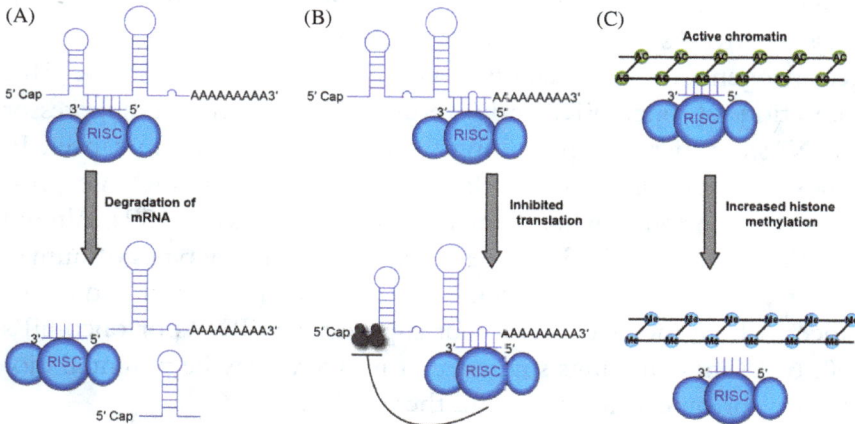

Figure 4-4. Mechanisms of post-transcriptional regulation of miRNA. (A) In the event of perfect pairing between miRNA and the target mRNA, miRNA leads to degradation of mRNA. (B) When there is no exact matching between miRNA and the target mRNA, miRNA may inhibit translation. (C) MiRNA may transcriptionally regulate the expression of a target by modifying histone modifications and chromatin remodeling pattern.

pairing between mRNA and miRNA is complete, it is most likely that degradation of the mRNA follows due to decreased steric hindrance. Secondly, by incomplete binding to mRNA, miRNA can inhibit translation. Thirdly, miRNA may regulate the expression of a target by modifying histone modifications and chromatin remodeling pattern. It should be noted that in some low-chance cases, some miRNAs can contrarily activate translation of their target mRNA when the cell is quiescent (not dividing or not preparing to divide).

In 2002, it was shown for the first time that miRNAs were involved in cancer [57]. MiRNAs regulate multiple biological processes including proliferation, cell death, development, and genomic stability — all essential for tumor development [6]. Calin *et al.* discovered that miR-15a and miR-16-1 were located in a region that was frequently lost in leukemia patients and that both miRNAs were deleted or significantly downregulated in almost 70% of all chronic lymphocytic leukemia patients [57]. Since that initial discovery, cancer-associated miRNAs are classified as either oncogenic miRNAs (oncomiR) or tumor-suppressive miRNAs. These miRNAs are usually located in cancer-associated gene regions. Whereas oncomiRs are frequently upregulated in cancer, where they target tumor suppressor genes for degradation and promote cancer cell growth; tumor-suppressor miRNAs are usually downregulated in cancer, since they target oncogenes for degradation and have an anti-tumor function [58]. Inhibition of oncomiRs and overexpression of tumor suppressor miRNAs are therefore promising for targeted therapies in cancer. In almost all stages of cancer development, dysregulated miRNA expression has been found when compared to normal tissues [59]. Altered miRNA expression profiles have been found in many types of human cancers, including colon cancer, brain tumors, lung cancer and breast cancer, where they work as tumor suppressor miRNAs or oncomiRs [60, 61]. These findings suggest that miRNAs may be potential biomarkers for cancer detection and therapy [52, 62, 63].

Compared to mRNA, analysis of miRNAs offers several advantages: (1) they are small and therefore more stable; (2) they can be extracted from frozen tissues, formalin-fixed paraffin-embedded tissues as well as blood, with little or no degradation [6].

4.4.1.1 *MicroRNAs in TNBC*

The advantages mentioned above make analysis of miRNAs ideal for clinical application, which is under active investigation in TNBC diagnosis and treatment [6]. Over the last 10 years, there have been multiple studies identifying miRNA changes associated with TNBC (**Table 4-1**). Scientists summarized the latest miRNA profiling, functional and prognostic findings that have been implicated in the pathology of TNBC. A number of miRNAs have been identified and validated that target key genes involved in critical cellular functions. As an example, the miR-200 family targets ZEB1/ZEB2, Suz 12, EphA2, MSN, FN1, TrkB, XIAP, all of which are important for cell proliferation, invasion, and migration [64–66]. Multiple studies have revealed that various miRNAs specifically target the three receptors ER, PR, and HER2 that become missing in TNBC development. In addition, the breast cancer susceptibility gene BRCA1 is also a target of miRNAs. The study of prognostic miRNAs by Liu *et al.* identified a signature of four miRNAs, i.e., miR-374b-5p ↑, miR-218-5p ↑, miR-126-3p ↑, miR-27b-3p ↓, that appeared to be associated with good prognosis in TNBC [67]. Despite these encouraging findings, there remains a need for better validation and reliability in the experimental conditions and subsequent analysis to define specific miRNAs as biomarkers of the disease [6].

Medimegh *et al.* have explored the expression levels of miRNAs in TNBC in comparison with non-TNBC cases and have found that miR-21, miR-146a, and miR-182 are significantly expressed in TNBC and miR-10b, miR-21, and miR-182 are significantly associated with lymph node metastasis in TNBC [68]. In addition, Cao *et al.* found that high expression of miR-454 is associated with poor prognosis in TNBC [69]. Furthermore, induction of some miRNAs such as miR-181a by genotoxic treatments may enhance TNBC survival and metastasis [70], which means antagonizing miRNAs may serve as a strategy to sensitize TNBC cells to chemotherapy.

As will be discussed in the next chapter, novel biomarkers or treatment targets are urgently required to improve disease outcomes of TNBC. MiRNAs represent attractive candidates for targeted therapies against TNBC, due to their natural ability to act as antisense

Table 4-1. Summary of miRNAs associated with TNBC.

	MicroRNA	Validated miRNA targets	Main biological function(s)
Down-regulation of onco-suppressor miRNAs	miR-200 family	ZEB1/ZEB2, Suz12, EphA2	Stimulation of differentiation & inhibition of EMT
	miR-205	E2F1, LAMC1	Reduction of proliferation, cell cycle
	miR-203	BIRC5, LASP1	Reduction of proliferation and inhibition of migration
	miR-31	WAVE3, RhoA; Radexin, PRKCE	Reduction of metastatic potential; induction of apoptosis and enhancement of chemosensitivity
	miR-34a	AXL, NOTCH1, TWIST1, ZEB	Impairment of migration. Inhibition of EMT signaling pathway
	miR-34c	MAP3K2	Inhibition of migration, invasion and EMT
	miR-145	ARF6, Mucin1	Inhibition of EMT and metastasis
	miR-139-5p	HRAS, NFKB1, PIK3CA, RAF, RHOT	Reduction of metastatic potential
	miR-193b	u-PA	Inhibition of cell invasion
	miR-335	SOX4, extracellular matrix	Metastasis suppressor
	miR-126	StarD10	Metastasis suppressor
	miR-17-5p	PDCD4, PTEN	Inhibition of PDCD4 or PTEN
	miR-455-3p	SMAD2, LTBR, EI24	Enhancement of cell proliferative, invasive and migration abilities
	miR-211-5p	SETBP1	Suppression of proliferation, invasion, migration and metastasis
Upregulation of oncogenic miRNAs	miR-181a	Bim, ATM	Inhibition of anoikisis, impairment of DNA double-strand-breaks repair
	miR-146	BRCA1	Control of BRCA1-mediated proliferation and homologous recombination
	miR-182	PFN1, FOXF2	Inhibition of cell proliferation and invasion & induction of apoptosis
	miR-221		Repression of E-cadherin
	miR-221	DNMT3b	Promotion of stemness of breast cancer cells
	miR-17, miR-20a		Decrease of TIMP2/3 expressions
	miR-10b	HOXD10	Metastasis inducer

Table 4-1. (*Continued*)

MicroRNA	Validated miRNA targets	Main biological function(s)
miR-1207	STAT6	Increase of cell proliferation
miR-301	FOXF2, BBC3, PTEN, COL2A1	Decrease of cell proliferation, clonogenicity, migration, invasion, tamoxifen resistance
miR-103/107	Dicer	Metastasis inducer
miR-199a	LCOR	Enhancement of cancer stem cell properties, protection from interferon signaling
miR-9	CDH1	Increase of cell motility and invasiveness

Ref: *Int J Mol Sci. 2015; 16(12).28347–28376.*

interactors and regulators of entire gene sets involved in malignancy and their superiority over mRNA profiling to accurately classify disease [6, 71].

4.4.2 Long Non-coding RNAs

Long non-coding RNAs (lncRNAs) are transcripts longer than 200 nt in length that are known not to produce any protein product. LncRNAs regulate target gene expression in various ways: epigenetic regulation, transcriptional regulation, and post-transcriptional regulation [45].

The expression of lncRNA possesses tissue- and cell type-specificity and is regulated during growth. Structurally, lncRNAs are similar to mRNAs, have many different types of transcripts and can translate into gene-encoding antisense transcripts. LncRNAs have vital regulatory functions and are closely linked to the progression of diseases. LncRNAs play their role by binding to DNA/RNA or protein. Some lncRNAs are actually precursors of other regulatory RNAs such as miRNAs or piRNAs. In contrast to miRNAs, lncRNAs do not have a general function per se, but instead, regulate gene expression and protein synthesis through various ways. Some lncRNAs are involved in the basic process of gene regulation, including chromatin modification and direct transcriptional regulation. Different forms of lncRNAs function via cis- or trans-regulation.

LncRNAs are emerging as important players in shifting the cancer-inducing paradigm. Altered expression of lncRNAs is specifically associated with tumorigenesis, tumor progression, and metastasis. They are easily, rapidly, and cost-effectively determined in tissues, serum and gastric juice, where potential features of ncRNAs make them highly versatile for analyses [45].

4.4.2.1 *LncRNAs in TNBC*

A large number of studies have shown evidence that the abnormal expression of lncRNA can induce many diseases, including cancer. The abnormal expression of lncRNA in tumor tissues has a wide distribution and may be involved in many types of malignancies, including leukemia, liver cancer, lung cancer, colorectal cancer, prostate cancer, and breast cancer.

Studies have begun reporting the relationship between lncRNAs and tumor metastasis and invasion in recent years. As described in Chapter 2, TNBC is a heterogeneous disease and even though a high number of targeted therapies have been clinically tested, this has not yet translated into a substantial clinical benefit for TNBC patients. Hence, it is necessary to identify highly sensitive biomarkers for a better stratification and treatment of these patients. Recently, lncRNAs have been reported to drive many important cancer phenotypes through their interactions with other cellular macromolecules [72, 73]. To date, it has been strongly proposed that a deeper functional understanding of lncRNAs will provide novel insights into the molecular mechanism of cancer. As such, lncRNAs are likely to serve as the basis for many clinical applications in oncology [74, 75].

As mentioned earlier, lncRNAs are regulatory non-coding RNAs rather than house-keeping RNAs. New evidence supports that lncRNAs have potential, diverse, and deep functional roles at the nuclear level, which include acting as a positive (activation) mechanism of transcriptional regulation, as well as inactivating epigenetic mechanisms, for example, X-chromosome inactivation, heterochromatin conformation, telomere maintenance, and pluripotency capacity modulation. Recently, accumulating evidence indicates that there is aberrant expression of lncRNAs in many cancer types including breast cancer. Lv *et al.* found that the expression of lncRNAs ANRIL, HIF1A-AS2, and UCA1 was significantly increased in the plasma of patients with TNBC [76], suggesting their potential use as TNBC-specific diagnostic biomarkers and/or molecular prognostic predictors.

In 2015, Shen *et al.* identified 1,758 lncRNAs and 1,254 mRNAs with significant expression differences in TNBC vs. normal adjacent tissues based on microarray analysis [77]. Subsequently, Yang *et al.* have been working on the identification and validation of the differential expression of lncRNAs by RNA-sequencing (RNA-seq) [74]. Some lncRNAs have been proposed as competitive endogenous RNA (ceRNA) (discussed below) for short ncRNA. LincRNA-RoR (regulator of reprogramming) is upregulated in pluripotent cells, where it functions as a ceRNA for miR-145, thereby protecting pluripotency factors from miR-mediated silencing, leading to loss of mature miR-145 expression. Recently, Eades *et al.* found that in TNBC, loss of miR-145 promoted tumor cell invasion. This is mediated via overexpression of ARF6, a protein implicated in tumor invasion, through disturbance of cell-cell adhesion by endocytosis of E-cadherin. In this case, lincRNA-RoR generates a competitive inhibition of miR-145, which alters ARF6 expression [78] (**Figure 4-5A**). The authors also reported overexpression of lincRNA-RoR in lymph node positive tumors of TNBC patients and reported the first ceRNA network in human cancer.

The expression of other lncRNAs, such as HOTAIR, has been shown to enhance the growth and metastasis in xenograft mammary tumors. Metastasis-associated lung adenocarcinoma transcript 1 (MALAT1) was found upregulated in TNBC tissues and is associated with tumor growth and metastasis, as well as poor OS in breast cancer. Downregulation of MALAT1 increased the expression of miR-1, while overexpression of miR-1 decreased MALAT1 expression. In this sense, MALAT1 exerts its function through the miR-1/Slug axis

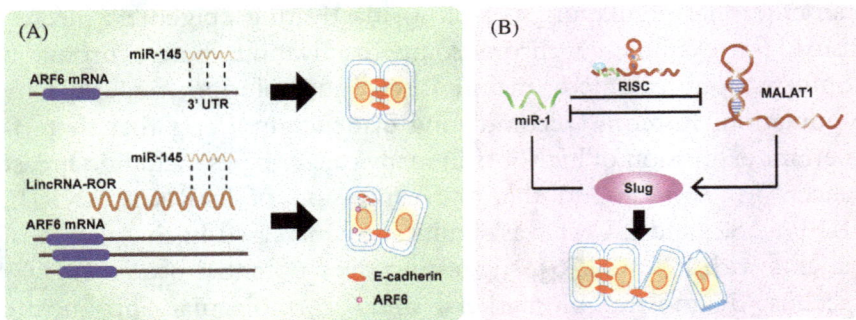

Figure 4-5. A model deciphering the molecular mechanism for lncRNAs involved in the tumorigenesis of human TNBC. (A) LincRNA-RoR as a miR-145 inhibitor (oncogene miRNA). (B) MALAT1 as a competitive endogenous RNA of miR-1 (tumor suppressor miRNA). ARF6, ADP-ribosylation factor 6; UTR, untranslated region; RISC, RNA-induced silencing complex; MALAT1, metastasis-associated lung adenocarcinoma transcript 1.

and thus MALAT1 may be a target for TNBC therapy [75] (**Figure 4-5B**). Recently, Lin *et al.* showed that the long intergenic non-coding RNA for kinase activation (LINK-A) was critical for growth factor-induced normoxic signaling pathway by recruiting breast tumor kinase (BRK) activated together with leucine-rich repeat kinase 2 (LRRK2) [79].

4.4.2.2 *Potential clinical applications of lncRNAs in TNBC*

LncRNAs play several roles in TNBC, but their biological participation is not yet fully understood. Some important advances have been reached, such as the study by Wang *et al.*, which described different expression patterns of lncRNAs in TNBC vs. non-cancer tissue. This may open new avenues for functional studies on lncRNAs that have not yet been totally defined as modulators of mRNA coding genes [80]. The lack of complete patterns impedes the development of new TNBC molecular targets as well as new targeted drugs, which could specifically target functional lncRNAs. Despite of this, however, lncRNA-based therapy would be a fascinating and novel therapeutic strategy. On that note, recently, Xia *et al.* designed an oligonucleotide with some chemical modifications which improved its half-life in serum. This molecule antagonizes the function of one tumorigenic lncRNA named ASBEL [81]. In this regard, they have proposed it as a new field of research of potential therapeutic tools for the treatment of TNBC.

Notably, lncRNAs could be detected in human body fluids, acting as biomarkers. Chen *et al.* provided useful information for exploring potential therapeutic targets for TNBC [82]. LncRNA expression could be regulated by conventional chemotherapy agents like receptor tyrosine kinases (RTKs) and non-RTKs by targeting multiple genes at once through unknown mechanisms. LncRNAs as biomarkers and their associated genetic-epigenetic and transcriptional mechanisms in co-expression patterns of mRNA coding genes open new insights for gene expression control and epigenetic events that could explain pathophysiology and/or pharmacological actions for clinical diagnosis, treatment response and prognosis of TNBC patients.

4.4.3 *Circular RNAs*

Circular RNAs (circRNAs) are a class of ncRNAs that are widely expressed in mammals. Plenty of circRNAs have been identified, but their potential functions are poorly understood. There are currently few reports describing the role of circRNAs in breast cancer. Liang *et al.* reported that circDENND4C was a HIF1α-associated circRNA that promoted the proliferation of breast cancer under hypoxia [83]. However, the function of circRNAs in TNBC progression is unclear. Revealing the role of circRNAs will be critical for understanding TNBC pathogenesis and offering a novel insight into identifying new biomarkers or therapeutic targets of TNBC.

It was reported that RNAs could act as ceRNAs to co-regulate each other by competing for shared miRNAs [84, 85]. Messenger RNAs, pseudogenes, lncRNAs, and circRNAs may all serve as ceRNAs. A number of findings indicate that circRNAs could function as miRNA sponges to contribute to the regulation of cancers.

Microarray analysis and qRT-PCR verified a circRNA termed circGFRA1 that was upregulated in TNBC. Kaplan-Meier survival analysis showed that upregulated circGFRA1 was correlated with poorer survival. Knockdown of circGFRA1 inhibited proliferation and promoted apoptosis in TNBC. He *et al.* showed that circGFRA1 acted as a ceRNA in TNBC by regulating miR-34a [86]. Via luciferase reporter assays, circGFRA1 was found to directly bind to miR-34a.

4.4.4   Interactions Between MicroRNAs and Other Epigenetic Mechanisms in TNBC

It has been shown that gene inactivation can be controlled by other epigenetic processes. Important proteins for the biogenesis of miRNAs can be methylated, leading to a decrease in the number of transcribed miRNAs. The majority of miRNAs have been found to be located in the intronic regions of protein-coding genes, and, as such, they can be co-regulated [87, 88]. Nevertheless, miRNAs also have their own promoters, which can be near/within CpG islands within the same intron. A study by Wee *et al.* identified that approximately 60% of 93 breast cancer-associated miRNAs are within 5 kb of a CpG island [89]. This suggests that miRNAs can be transcribed from their own promoters and that these promoters might be regulated by DNA methylation.

Lehmann *et al.* have shown that miR-9-1, miR-124a-3, miR-148, miR-152, and miR-663 are epigenetically inactivated through hypermethylation in breast cancer [90]. Furthermore, scientists identified promoter hypermethylation as one of the major mechanisms for silencing miR-31 in breast cancer and in TNBC cell lines. MiR-31 maps to the intronic sequence of a novel long non-coding RNA (lncRNA), LOC554202 and the regulation of its transcriptional activity is under control of LOC554202. Both miR-31 and LOC554202 are downregulated in TNBC cell lines of basal subtype and overexpressed in the luminal counterpart. Treatment of TNBC cell lines with either a demethylating agent alone or in combination with a deacetylating agent resulted in a significant increase of both miR-31 and its host gene, suggesting an epigenetic mechanism for the silencing of these two genes by promoter hypermethylation. Both methylation-specific PCR and sequencing of bisulfite-converted DNA demonstrated that the LOC554202 promoter-associated CpG island is heavily methylated in TNBC cell lines and hypomethylated in the luminal subtypes [91]. Additionally, miR-31 hypermethylation in TNBC leads to an increase in the expression of its pro-metastatic target genes (RhoA and WAVE3) [91]. As mentioned above, miRNAs can also control the epigenetic machinery.

MiRNAs have been shown to target DNMT enzymes and influence the DNA methylation process. Fabbri *et al.* were the first to identify that the miR-29 family directly targets DNMT3a and DNMT3b [92]. This also causes reexpression of methylation-silenced tumor suppressor genes (e.g., FHIT, WWOX) [92]. Later, these miRNAs were referred to as epi-miRs [90]. The miR-148 family can also target DNMT3b, resulting in a decrease in DNA methylation levels and altered splicing of DNMT3b [93]. Interestingly, miR-148a is also epigenetically regulated through promoter hypermethylation, suggesting an epigenetic feedback loop [94]. In summary, a better understanding of the interactions between miRNAs and other epigenetic control mechanisms will improve the knowledge of cancer development and progression, which will lead to improved diagnostic and prognostic markers [6].

References

1. Waddington CH. (2012). The epigenotype. 1942. *Int J Epidemiol,* **41**(1): 10–13.
2. Holliday R. (1987). The inheritance of epigenetic defects. *Science,* **238**(4824): 163–170.
3. Abdel-Hafiz HA, Horwitz KB. (2015). Role of epigenetic modifications in luminal breast cancer. *Epigenomics,* 7(5): 847–862.
4. Virani S, Colacino JA, Kim JH, Rozek LS. (2012). Cancer epigenetics: a brief review. *ILAR J,* **53**(3–4): 359–369.
5. Evron E, Dooley WC, Umbricht CB, Rosenthal D, Sacchi N, Gabrielson E, Soito AB, Hung DT, Ljung B, Davidson NE *et al.* (2001). Detection of breast cancer cells in ductal lavage fluid by methylation-specific PCR. *Lancet,* **357**(9265): 1335–1336.
6. Mathe A, Scott RJ, Avery-Kiejda KA. (2015). MiRNAs and Other Epigenetic Changes as Biomarkers in Triple Negative Breast Cancer. *Int J Mol Sci,* **16**(12): 28347–28376.

7. Moore LD, Le T, Fan G. (2013). DNA methylation and its basic function. *Neuropsychopharmacology*, **38**(1): 23–38.

8. Straussman R, Nejman D, Roberts D, Steinfeld I, Blum B, Benvenisty N, Simon I, Yakhini Z, Cedar H. (2009). Developmental programming of CpG island methylation profiles in the human genome. *Nat Struct Mol Biol*, **16**(5): 564–571.

9. Li E, Bestor TH, Jaenisch R. (1992). Targeted mutation of the DNA methyltransferase gene results in embryonic lethality. *Cell*, **69**(6): 915–926.

10. Okano M, Bell DW, Haber DA, Li E. (1999). DNA methyltransferases Dnmt3a and Dnmt3b are essential for de novo methylation and mammalian development. *Cell*, **99**(3): 247–257.

11. Lopez-Serra L, Esteller M. (2008). Proteins that bind methylated DNA and human cancer: reading the wrong words. *Br J Cancer*, **98**(12): 1881–1885.

12. Medvedeva YA, Khamis AM, Kulakovskiy IV, Ba-Alawi W, Bhuyan MS, Kawaji H, Lassmann T, Harbers M, Forrest AR, Bajic VB *et al.* (2014). Effects of cytosine methylation on transcription factor binding sites. *BMC Genomics*, **15**: 119.

13. Watt F, Molloy PL. (1988). Cytosine methylation prevents binding to DNA of a HeLa cell transcription factor required for optimal expression of the adenovirus major late promoter. *Genes Dev*, **2**(9): 1136–1143.

14. Supic G ZK, and Magic Z. (2018). Chapter 5: Epigenetic Therapy of Cancer — From Mechanisms to Clinical Utility, in book: Epigenetics in cancer, Editors: Fu J, Imani S. *Narosa Publishing House/Science Press* 2018.

15. Imperiale TF, Ransohoff DF, Itzkowitz SH, Levin TR, Lavin P, Lidgard GP, Ahlquist DA, Berger BM. (2014). Multitarget stool DNA testing for colorectal-cancer screening. *N Engl J Med*, **370**(14): 1287–1297.

16. Stirzaker C, Zotenko E, Clark SJ. (2016). Genome-wide DNA methylation profiling in triple-negative breast cancer reveals epigenetic signatures with important clinical value. *Mol Cell Oncol*, **3**(1): e1038424.

17. Stirzaker C, Zotenko E, Song JZ, Qu W, Nair SS, Locke WJ, Stone A, Armstong NJ, Robinson MD, Dobrovic A *et al.* (2015). Methylome sequencing in triple-negative breast cancer reveals distinct methylation clusters with prognostic value. *Nat Commun*, **6**: 5899.

18. Shanmugam MK, Arfuso F, Arumugam S, Chinnathambi A, Bian J, Warrier S, Wang LZ, Kumar AP, Ahn KS, Sethi G *et al.* (2017). Role of novel histone modifications in cancer. *Oncotarget*, **9**(13): 11414–11426.

19. Yang XJ. (2004). The diverse superfamily of lysine acetyltransferases and their roles in leukemia and other diseases. *Nucleic Acids Res,* **32**(3): 959–976.

20. Bustos MA, Salomon MP, Nelson N, Hsu SC, DiNome ML, Hoon DS, Marzese DM. (2017). Genome-wide chromatin accessibility, DNA methylation and gene expression analysis of histone deacetylase inhibition in triple-negative breast cancer. *Genom Data,* **12**: 14–16.

21. Ono H, Sowa Y, Horinaka M, Iizumi Y, Watanabe M, Morita M, Nishimoto E, Taguchi T, Sakai T. (2018). The histone deacetylase inhibitor OBP-801 and eribulin synergistically inhibit the growth of triple-negative breast cancer cells with the suppression of survivin, Bcl-xL, and the MAPK pathway. *Breast Cancer Res Treat,* **171**(1): 43–52.

22. D'Souza W SD. (2018). Chapter 2: DNA Methylation and Cancer, in book: Epigenetics in cancer, Editors: Fu J, Imani S. *Narosa Publishing House/Science Press.*

23. Fagone P, Jackowski S. (2013). Phosphatidylcholine and the CDP-choline cycle. *Biochim Biophys Acta,* **1831**(3): 523–532.

24. Vance JE, Vance DE. (2004). Phospholipid biosynthesis in mammalian cells. *Biochem Cell Biol,* **82**(1): 113–128.

25. Ridgway ND. (2013). The role of phosphatidylcholine and choline metabolites to cell proliferation and survival. *Crit Rev Biochem Mol Biol,* **48**(1): 20–38.

26. Podo F, Canevari S, Canese R, Pisanu ME, Ricci A, Iorio E. (2011). MR evaluation of response to targeted treatment in cancer cells. *NMR Biomed,* **24**(6): 648–672.

27. Podo F, Paris L, Cecchetti S, Spadaro F, Abalsamo L, Ramoni C, Ricci A, Pisanu ME, Sardanelli F, Canese R *et al.* (2016). Activation of Phosphatidylcholine-Specific Phospholipase C in Breast and Ovarian Cancer: Impact on MRS-Detected Choline Metabolic Profile and Perspectives for Targeted Therapy. *Front Oncol,* **6**: 171–178.

28. Beloueche-Babari M, Chung YL, Al-Saffar NM, Falck-Miniotis M, Leach MO. (2010). Metabolic assessment of the action of targeted cancer therapeutics using magnetic resonance spectroscopy. *Br J Cancer,* **102**(1): 1–7.

29. Glunde K, Bhujwalla ZM, Ronen SM. (2011). Choline metabolism in malignant transformation. *Nat Rev Cancer,* **11**(12): 835–848.

30. Podo F. (1999). Tumour phospholipid metabolism. *NMR Biomed,* **12**(7): 413–439.

31. Iorio E, Caramujo MJ, Cecchetti S, Spadaro F, Carpinelli G, Canese R, Podo F. (2016). Key Players in Choline Metabolic Reprograming in Triple-Negative Breast Cancer. *Front Oncol*, 6: 205–212.

32. Thapa N, Choi S, Tan X, Wise T, Anderson RA. (2015). Phosphatidylinositol Phosphate 5-Kinase Igamma and Phosphoinositide 3-Kinase/Akt Signaling Couple to Promote Oncogenic Growth. *J Biol Chem*, 290(30): 18843–18854.

33. Paris L, Cecchetti S, Spadaro F, Abalsamo L, Lugini L, Pisanu ME, Iorio E, Natali PG, Ramoni C, Podo F. (2010). Inhibition of phosphatidyl-choline-specific phospholipase C downregulates HER2 overexpression on plasma membrane of breast cancer cells. *Breast Cancer Res*, 12(3): R27.

34. Miyake T, Parsons SJ. (2012). Functional interactions between Choline kinase alpha, epidermal growth factor receptor and c-Src in breast cancer cell proliferation. *Oncogene*, 31(11): 1431–1441.

35. Abalsamo L, Spadaro F, Bozzuto G, Paris L, Cecchetti S, Lugini L, Iorio E, Molinari A, Ramoni C, Podo F. (2012). Inhibition of phosphatidyl-choline-specific phospholipase C results in loss of mesenchymal traits in metastatic breast cancer cells. *Breast Cancer Res*, 14(2): R50.

36. Pisterzi P ML, Caramujo MJ, Iorio E, Podo F, Cecchetti S. (2015). Phosphatidylcholine-specific phospholipase C inhibition as a new thera-peutic approach to control triple-negative breast cancer cells proliferation. *Proceedings of the Special Conference EACR-AACR-SIC 2015*.

37. Aboagye EO, Bhujwalla ZM. (1999). Malignant transformation alters membrane choline phospholipid metabolism of human mammary epi-thelial cells. *Cancer Res*, 59(1): 80–84.

38. Glunde K, Jie C, Bhujwalla ZM. (2004). Molecular causes of the aber-rant choline phospholipid metabolism in breast cancer. *Cancer Res*, 64(12): 4270–4276.

39. Eliyahu G, Kreizman T, Degani H. (2007). Phosphocholine as a bio-marker of breast cancer: molecular and biochemical studies. *Int J Cancer*, 120(8): 1721–1730.

40. Beckmann MW, Niederacher D, Schnurch HG, Gusterson BA, Bender HG. (1997). Multistep carcinogenesis of breast cancer and tumour het-erogeneity. *J Mol Med (Berl)*, 75(6): 429–439.

41. ESMRMB, 32nd Annual Scientific Meeting, Edinburgh, UK, 1–3 October: EPOS Poster/Paper Poster/Clinical Review Poster/Software Exhibits. *MAGMA*, 28 Suppl 1: 419–519.

42. Gadiya M, Mori N, Cao MD, Mironchik Y, Kakkad S, Gribbestad IS, Glunde K, Krishnamachary B, Bhujwalla ZM. (2014). Phospholipase

D1 and choline kinase-alpha are interactive targets in breast cancer. *Cancer Biol Ther*, **15**(5): 593–601.

43. Mori N, Glunde K, Takagi T, Raman V, Bhujwalla ZM. (2007). Choline kinase down-regulation increases the effect of 5-fluorouracil in breast cancer cells. *Cancer Res*, **67**(23): 11284–11290.

44. Mori N, Wildes F, Kakkad S, Jacob D, Solaiyappan M, Glunde K, Bhujwalla ZM. (2015). Choline kinase-alpha protein and phosphatidyl-choline but not phosphocholine are required for breast cancer cell survival. *NMR Biomed*, **28**(12): 1697–1706.

45. Imani S ZX, Fu S, and Fu J. (2018). Chapter 3: Non-coding RNAs in Cancer, in book: Epigenetics in cancer, Editors: Fu J, Imani S. *Narosa Publishing House/Science Press*.

46. Mouh FZ, Mzibri ME, Slaoui M, Amrani M. (2016). Recent Progress in Triple Negative Breast Cancer Research. *Asian Pac J Cancer Prev*, **17**(4): 1595–1608.

47. Imani S, Wei C, Cheng J, Khan MA, Fu S, Yang L, Tania M, Zhang X, Xiao X, Zhang X *et al.* (2017). MicroRNA-34a targets epithelial to mesenchymal transition-inducing transcription factors (EMT-TFs) and inhibits breast cancer cell migration and invasion. *Oncotarget*, **8**(13): 21362–21379.

48. Lee RC, Feinbaum RL, Ambros V. (1993). The C. elegans heterochronic gene lin-4 encodes small RNAs with antisense complementarity to lin-14. *Cell*, **75**(5): 843–854.

49. O'Day E, Lal A. (). MicroRNAs and their target gene networks in breast cancer. *Breast Cancer Res*, **12**(2): 201–210.

50. Ying SY, Chang DC, Lin SL. (2008). The microRNA (miRNA): overview of the RNA genes that modulate gene function. *Mol Biotechnol*, **38**(3): 257–268.

51. Beezhold KJ, Castranova V, Chen F. (2010). Microprocessor of microRNAs: regulation and potential for therapeutic intervention. *Mol Cancer*, **9**: 134–142.

52. Calin GA, Croce CM. (2006). MicroRNA signatures in human cancers. *Nat Rev Cancer*, **6**(11): 857–866.

53. Ha M, Kim VN. (2014). Regulation of microRNA biogenesis. *Nat Rev Mol Cell Biol*, **15**(8): 509–524.

54. Kozomara A, Griffiths-Jones S. (2014). miRBase: annotating high confidence microRNAs using deep sequencing data. *Nucleic Acids Res*, **42**(Database issue): D68–73.

55. Andorfer CA, Necela BM, Thompson EA. Perez EA. (2011). MicroRNA signatures: clinical biomarkers for the diagnosis and treatment of breast cancer. *Trends Mol Med*, **17**(6): 313–319.

56. Friedman RC, Farh KK, Burge CB, Bartel DP. (2009). Most mammalian mRNAs are conserved targets of microRNAs. *Genome Res*, 19(1): 92–105.

57. Calin GA, Dumitru CD, Shimizu M, Bichi R, Zupo S, Noch E, Aldler H, Rattan S, Keating M, Rai K *et al.* (2002). Frequent deletions and down-regulation of micro- RNA genes miR15 and miR16 at 13q14 in chronic lymphocytic leukemia. *Proc Natl Acad Sci USA*, 99(24): 15524–15529.

58. Voorhoeve PM. (2010). MicroRNAs: Oncogenes, tumor suppressors or master regulators of cancer heterogeneity? *Biochim Biophys Acta*, 1805(1): 72–86.

59. Izumiya M, Tsuchiya N, Okamoto K, Nakagama H. (2011). Systematic exploration of cancer-associated microRNA through functional screening assays. *Cancer Sci*, 102(9): 1615–1621.

60. Baffa R, Fassan M, Volinia S, O'Hara B, Liu CG, Palazzo JP, Gardiman M, Rugge M, Gomella LG, Croce CM *et al.* (2009). MicroRNA expression profiling of human metastatic cancers identifies cancer gene targets. *J Pathol*, 219(2): 214–221.

61. Di Leva G, Briskin D, Croce CM. (2012). MicroRNA in cancer: new hopes for antineoplastic chemotherapy. *Ups J Med Sci*, 117(2): 202–216.

62. Heneghan HM, Miller N, Lowery AJ, Sweeney KJ, Kerin MJ. (2009). MicroRNAs as Novel Biomarkers for Breast Cancer. *J Oncol*, 2009: 950201.

63. Jeffrey SS. (2008). Cancer biomarker profiling with microRNAs. *Nat Biotechnol*, 26(4): 400–401.

64. Aydogdu E, Katchy A, Tsouko E, Lin CY, Haldosen LA, Helguero L, Williams C. (2012). MicroRNA-regulated gene networks during mammary cell differentiation are associated with breast cancer. *Carcinogenesis*, 33(8): 1502–1511.

65. Howe EN, Cochrane DR, Richer JK. (2011). Targets of miR-200c mediate suppression of cell motility and anoikis resistance. *Breast Cancer Res*, 13(2): R45.

66. Ren Y, Han X, Yu K, Sun S, Zhen L, Li Z, Wang S. (2014). microRNA-200c downregulates XIAP expression to suppress proliferation and promote apoptosis of triple-negative breast cancer cells. *Mol Med Rep*, 10(1): 315–321.

67. Liu Y, Cai Q, Bao PP, Su Y, Cai H, Wu J, Ye F, Guo X, Zheng W, Zheng Y *et al.* (2015). Tumor tissue microRNA expression in association with triple-negative breast cancer outcomes. *Breast Cancer Res Treat*, 152(1): 183–191.

68. Medimegh I, Omrane I, Privat M, Uhrhummer N, Ayari H, Belaiba F, Benayed F, Benromdhan K, Mader S, Bignon IJ *et al.* (2014). MicroRNAs expression in triple negative vs non triple negative breast cancer in Tunisia: interaction with clinical outcome. *PloS One*, 9(11): e111877.

69. Cao ZG, Li JJ, Yao L, Huang YN, Liu YR, Hu X, Song CG, Shao ZM. (2016). High expression of microRNA-454 is associated with poor prognosis in triple-negative breast cancer. *Oncotarget*, 7(40): 64900–64909.

70. Niu J, Xue A, Chi Y, Xue J, Wang W, Zhao Z, Fan M, Yang CH, Shao ZM, Pfeffer LM *et al.* (2016). Induction of miRNA-181a by genotoxic treatments promotes chemotherapeutic resistance and metastasis in breast cancer. *Oncogene*, 35(10): 1302–1313.

71. McGuire A, Brown JA, Kerin MJ. (2015). Metastatic breast cancer: the potential of miRNA for diagnosis and treatment monitoring. *Cancer Metastasis Rev*, 34(1): 145–155.

72. Huarte M. (2015). The emerging role of lncRNAs in cancer. *Nat Med*, 21(11): 1253–1261.

73. Evans JR, Feng FY, Chinnaiyan AM. (2016). The bright side of dark matter: lncRNAs in cancer. *J Clin Invest*, 126(8): 2775–2782.

74. Yang F, Liu YH, Dong SY, Yao ZH, Lv L, Ma RM, Dai XX, Wang J, Zhang XH, Wang OC. (2016). Co-expression networks revealed potential core lncRNAs in the triple-negative breast cancer. *Gene*, 591(2): 471–477.

75. Rodriguez Bautista R, Ortega Gomez A, Hidalgo Miranda A, Zentella Dehesa A, Villarreal-Garza C, Avila-Moreno F, Arrieta O. (2018). Long non-coding RNAs: implications in targeted diagnoses, prognosis, and improved therapeutic strategies in human non- and triple-negative breast cancer. *Clin Epigenetics*, 10: 88–102.

76. Liu M, Xing LQ, Liu YJ. (2017). A three-long noncoding RNA signature as a diagnostic biomarker for differentiating between triple-negative and non-triple-negative breast cancers. *Medicine (Baltimore)*, 96(9): e6222.

77. Shen X, Xie B, Ma Z, Yu W, Wang W, Xu D, Yan X, Chen B, Yu L, Li J *et al.* (2015). Identification of novel long non-coding RNAs in triple-negative breast cancer. *Oncotarget*, 6(25): 21730–21739.

78. Eades G, Wolfson B, Zhang Y, Li Q, Yao Y, Zhou Q. (2015). lincRNA-RoR and miR-145 regulate invasion in triple-negative breast cancer via targeting ARF6. *Mol Cancer Res*, 13(2): 330–338.

79. Lin A, Li C, Xing Z, Hu Q, Liang K, Han L, Wang C, Hawke DH, Wang S, Zhang Y *et al.* (2016). The LINK-A lncRNA activates nor-

moxic HIF1alpha signalling in triple-negative breast cancer. *Nat Cell Biol*, **18**(2): 213–224.

80. Wang YL, Overstreet AM, Chen MS, Wang J, Zhao HJ, Ho PC, Smith M, Wang SC. (2015). Combined inhibition of EGFR and c-ABL suppresses the growth of triple-negative breast cancer growth through inhibition of HOTAIR. *Oncotarget*, **6**(13): 11150–11161.

81. Xia Y, Xiao X, Deng X, Zhang F, Zhang X, Hu Q, Sheng W. (2017). Targeting long non-coding RNA ASBEL with oligonucleotide antagonist for breast cancer therapy. *Cancer Biol Ther*, **489**(4): 386–392.

82. Chen C, Li Z, Yang Y, Xiang T, Song W, Liu S. (2015). Microarray expression profiling of dysregulated long non-coding RNAs in triple-negative breast cancer. *Cancer Biol Ther*, **16**(6): 856–865.

83. Liang G, Liu Z, Tan L, Su AN, Jiang WG, Gong C. (2017). HIF1alpha-associated circDENND4C Promotes Proliferation of Breast Cancer Cells in Hypoxic Environment. *Anticancer Res*, **37**(8): 4337–4343.

84. Tay Y, Rinn J, Pandolfi PP. (2014). The multilayered complexity of ceRNA crosstalk and competition. *Nature*, **505**(7483): 344–352.

85. Salmena L, Poliseno L, Tay Y, Kats L, Pandolfi PP. (2011). A ceRNA hypothesis: the Rosetta Stone of a hidden RNA language? *Cell*, **146**(3): 353–358.

86. He R, Liu P, Xie X, Zhou Y, Liao Q, Xiong W, Li X, Li G, Zeng Z, Tang H. (2017). circGFRA1 and GFRA1 act as ceRNAs in triple negative breast cancer by regulating miR-34a. *J Exp Clin Cancer Res*, **36**(1): 145–156.

87. Rodriguez A, Griffiths-Jones S, Ashurst JL, Bradley A. (2004). Identification of mammalian microRNA host genes and transcription units. *Genome Res*, **14**(10A): 1902–1910.

88. Ying SY, Lin SL. (2005). Intronic microRNAs. *Biochem Biophys Res Commun*, **326**(3): 515–520.

89. Wee EJ, Peters K, Nair SS, Hulf T, Stein S, Wagner S, Bailey P, Lee SY, Qu WJ, Brewster B *et al.* (2012). Mapping the regulatory sequences controlling 93 breast cancer-associated miRNA genes leads to the identification of two functional promoters of the Hsa-mir-200b cluster, methylation of which is associated with metastasis or hormone receptor status in advanced breast cancer. *Oncogene*, **31**(38): 4182–4195.

90. Fabbri M, Calin GA. (2010). Epigenetics and miRNAs in human cancer. *Adv Genet*, **70**: 87–99.

91. Augoff K, McCue B, Plow EF, Sossey-Alaoui K. (2012). miR-31 and its host gene lncRNA LOC554202 are regulated by promoter hypermethylation in triple-negative breast cancer. *Mol Cancer*, **11**:5–17.

92. Fabbri M, Garzon R, Cimmino A, Liu Z, Zanesi N, Callegari E, Liu S, Alder H, Costinean S, Fernandez-Cymering C *et al.* (2007). MicroRNA-29 family reverts aberrant methylation in lung cancer by targeting DNA methyltransferases 3A and 3B. *Proc Natl Acad Sci USA*, **104**(40): 15805–15810.

93. Duursma AM, Kedde M, Schrier M, le Sage C, Agami R. (2008). miR-148 targets human DNMT3b protein coding region. *RNA*, **14**(5): 872–877.

94. Lehmann U, Hasemeier B, Christgen M, Muller M, Romermann D, Langer F, Kreipe H. (2008). Epigenetic inactivation of microRNA gene hsa-mir-9–1 in human breast cancer. *J Pathol*, **214**(1): 17–24.

Chapter FIVE
Biomarkers of Triple-Negative Breast Cancer

Qiujun Lu[1], Mi Wu[2,3], Cuiyan Wu[1], Meiling Liu[1],
Faqing Tang[4], *and* Youyu Zhang[1,*]

Contents

*Corresponding author: Youyu Zhang, E-mail: zhangyy@hunnu.edu.cn
[1]Key Laboratory of Chemical Biology and Traditional Chinese Medicine Research (Ministry of Education), College of Chemistry and Chemical Engineering, Hunan Normal University, Changsha, China.
[2]Key Laboratory of Translational Cancer Stem Cell Research, Hunan Normal University, Changsha, Hunan, China.
[3]Departments of Pathology and Pathophysiology, Hunan Normal University School of Medicine, Changsha, Hunan, China.
[4]Hunan Cancer Hospital & The Affiliated Cancer Hospital of Xiangya School of Medicine, Central South University, Changsha, Hunan, China.

Triple-negative breast cancer (TNBC) is defined by the lack of clinically significant levels of estrogen receptor (ER), progesterone receptor (PR), and human epidermal growth factor receptor 2 (HER2) expression. It has different clinical and pathological features as compared to other subtypes of breast cancer. Moreover, patients with TNBC also have inferior disease-free survival (DFS) and overall survival (OS) [1]. Due to the lack of hormone receptor and HER2 expression, this special type of breast cancer cannot be treated with classical endocrine therapy and HER2-based targeted therapy [2]. Although preoperative chemotherapy is effective for some patients, TNBC still has the characteristics of higher relapse, higher invasive metastasis, and poorer prognosis compared to other subtypes of breast cancer [3–6]. Therefore, identification of biomarkers can be of particularly importance in directing the diagnosis and treatment decisions of TNBC. In this chapter, we mainly focus on the application of circulating tumor cells, circulating tumor DNA, and biomolecules as biomarkers in TNBC. The possible roles of epigenetic modulating molecules such as microRNA and long non-coding RNA as TNBC biomarkers are discussed in Chapter 4.

5.1 Circulating Tumor Cells in TNBC

Circulating tumor cells (CTCs) are cancer cells of solid tumor origin found in the peripheral blood [7, 8] (**Figure 5-1**). Most CTCs undergo apoptosis or phagocytosis after entering blood circulation. However, a few can escape and transfer to the distant organs, forming a metastatic focus eventually. Thus, the detection and analysis of CTCs have an important role in deciphering tumor

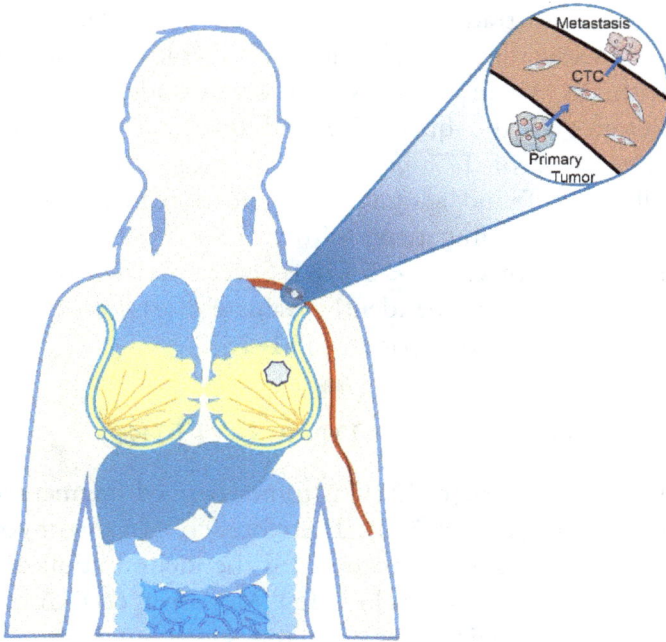

Figure 5-1. CTCs are tumor cells shed into the vasculature. CTCs are generated in the primary tumor site and move to a distant location via blood circulation.

dissemination, progression, and prognosis [9]. Among the commercial detection technologies for CTCs, the Veridex CellSearch™ System is US FDA-cleared for use in monitoring patients with metastatic breast, colorectal, and prostate cancers [10]. Detection of CTCs is achieved via epithelial antigens for immunomagnetic bead separation (antibody-based capture of CTCs, based on cell surface markers such as the epithelial cell adhesion molecule (EpCAM)) or on the physical properties of CTCs including filtration based on larger CTC size, differences in density and electrical properties, etc. Furthermore, molecular assays including immunocytochemistry, immunofluorescence, fluorescence in situ hybridization (FISH), and PCR-based techniques, coupled with CTC detection approaches allow molecular characterization of CTCs, for example, gene expression profiles, DNA methylation, mutations, and microRNA (miRNA) expression [10–12].

Compared with traditional imaging methods, CTCs can predict disease status earlier and in real time, and CTC level has a closer correlation with OS. There is a growing body of evidence for the prognostic relevance of the qualitative and quantitative assessment of CTCs. In a study of 177 patients with metastatic breast cancer, Cristofanilli *et al.* found a significant reduction in progression-free survival (PFS) and OS in patients with CTCs $\geq$ 5 per 7.5 ml of whole blood compared with CTCs < 5 [13]. Several studies indicated that CTCs can be used to independently evaluate disease progression and predict prognosis in TNBC patients [14–16].

5.2 Circulating Tumor DNA in TNBC

Circulating tumor DNA (ctDNA) is tumor-derived fragmented DNA in the bloodstream (**Figure 5-2**). Therefore, ctDNA might constitute potential sources for genetic material for the identification of tumor-associated molecular alterations [17, 18]. Moreover, ctDNA has a relatively short half-life ranging from 15 minutes to about 2 hours, and can therefore be used as a dynamic biomarker providing an accurate and possibly real-time monitoring of tumor development [19–21].

Various molecular methods, including PCR- and next-generation sequencing (NGS)-based techniques, have been developed to detect ctDNA from patient samples [21] (**Figure 5-3**).

The amount of ctDNA in the plasma is directly correlated to the tumor burden and the stage of the disease. Alterations of ctDNA have been studied in breast cancer as a surrogate for genetic and epigenetic alterations found in primary tumor tissues. Analysis of ctDNA has potential applications in the management of TNBC [21]. A series of studies of ctDNA in patients with primary breast cancer demonstrated that ctDNA monitoring is highly accurate for postsurgical discrimination between patients with and without eventual clinical recurrence, and the ctDNA-based detection is on an average of 11 months earlier than clinical detection of metastasis [22]. As far as TNBC is concerned, Madic *et al.* studied the ctDNA of 40 TNBC patients and found that ctDNA levels have no prognostic impact on time to progression [23]. These results indicated that the prognostic value of ctDNA might differ among breast cancer subgroups.

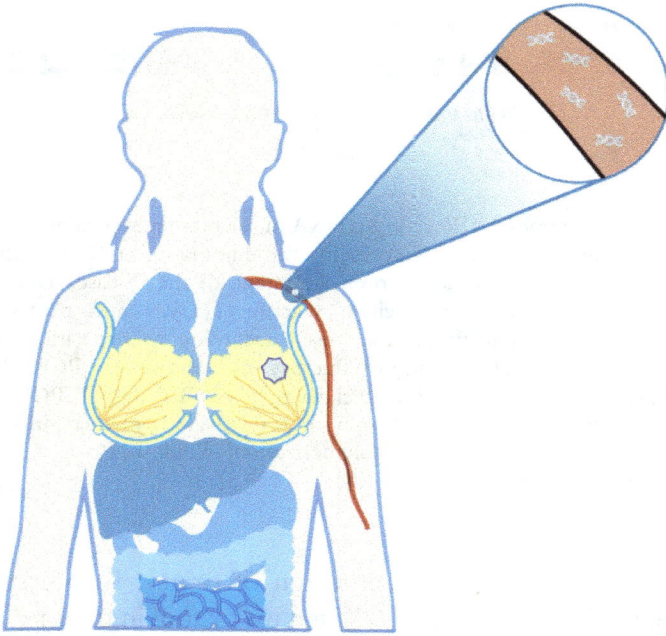

Figure 5-2. CtDNA intravasates from tumors into the bloodstream. CtDNA is fragmented DNA released from tumor cells into the bloodstream, which might provide genetic information for the identification of tumor-associated molecular alterations.

In addition, some studies have shown that ctDNA may be used to identify molecular alterations of therapeutic effect, and various suitable targets (such as fibroblast growth factor receptor (FGFR) [24–26], HER3 [27], PIK3CA [28], protein kinase B (Akt) [29], BRCA1/2 [30, 31], etc.) have already been identified and could be tracked in the plasma of TNBC patients.

5.3 Molecular Biomarkers in TNBC

5.3.1 *Cytokeratins*

Cytokeratins (CKs) are keratin proteins found in the cytoskeleton of epithelial cells. Different epithelial tissues express different CKs at the time of their terminal differentiation. Similarly, different cancers

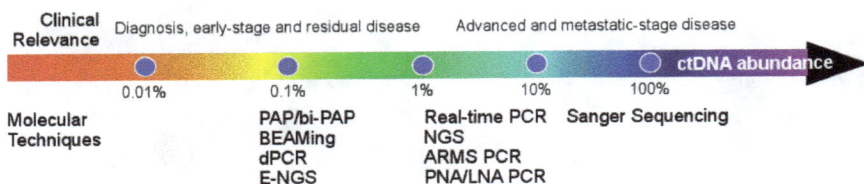

Figure 5-3. Techniques for detecting ctDNA in the plasma of patients and its clinical relevance. The colorful arrow represents the amount of ctDNA in the plasma directly correlating to the tumor burden and stage of the disease. The molecular techniques including polymerase chain reaction (PCR)- and next-generation sequencing (NGS)-based techniques are used according to the various detection thresholds. ARMS PCR, amplification-refractory mutation system PCR; BEAMing, beads, emulsion and amplification technology; dPCR, digital PCR; E-NGS, enhanced-NGS; PAP/bi-PAP, pyrophosphorolysis-activated polymerization/bidirectional pyrophosphorolysis-activated polymerization; PNA/LNA: peptide nucleic acid/locked nucleic acid.

express specific CKs of the epithelium of origin. In breast cancer, analysis of the level and type of CKs is a major tool in tumor diagnosis, providing molecular parameters to assess its differentiation status. In the early stages, CKs are used to distinguish malignant breast lesions from benign ones, but later studies found that CKs could be used for prognosis prediction [32, 33]. The expression of CK5/6, CK14, and CK17 is associated with poor prognosis in breast cancer [34–36].

As mentioned in Chapter 1, TNBC has higher CK levels, particularly CK5/6, than non-TNBC. Kim *et al.* found that basal-like TNBC with nodal and distant metastases is significantly associated with a higher intratumoral expression of CK5/6 and EGFR compared to those in the node-negative group. High level of intratumoral EGFR and CK5/6 expression may play a role in the development of nodal or distant metastases in the patients with basal-like TNBC tumors and may be predictive of metastatic disease [37]. Liu *et al.* analyzed the CK5/6 and CK17 of 112 TNBC patients by immunohistochemistry and showed that positive staining for CK5/6 or CK17 was associated with worse DFS, OS, high tumor grade, and positive axillary lymph nodes [38]. Thike *et al.* analyzed 653 TNBC patients, and the results also showed CK17 positivity impacted adversely on DFS and OS [39]. Since the CKs had prognostic implications on survival, possibilities exist for future targeted therapy for CKs in TNBC.

5.3.2 *EGFR*

Epidermal growth factor receptor (EGFR), also known as HER1, is a transmembrane protein with a relative molecular mass of 170 kDa. The EGFR family consists of four members, including EGFR (ErbB1), HER2 (ErbB2), HER3 (ErbB3), and HER4 (ErbB4) [40]. The EGFR family proteins are composed of three parts: extracellular region, transmembrane region, and intracellular region with tyrosine kinase activity. EGFR is widely distributed in human epidermal cells, stromal cells, some glial cells and smooth muscle cells, and is an important regulator of cell growth, differentiation, and survival. Although present in normal cells, EGFR is overexpressed in a variety of cancers, such as breast cancer, head-and-neck cancer, non-small-cell lung cancer, renal cancer, ovarian cancer, and colon cancer [41]. Such overexpression produces intense signal transduction and activation of downstream signaling pathways, resulting in more aggressive growth and invasiveness characteristics.

Compared with other subtypes of breast cancer, TNBC is more likely to express EGFR (discussed in Chapter 1). This has further been confirmed by the fact that there is a negative correlation between EGFR and hormone receptor status in breast cancer. EGFR expression is correlated with a less favorable response to chemotherapy and poorer survival [42]. In a study of 284 TNBC patients, it was found that 57.4% of patients showed EGFR expression and that patients with EGFR expression had worse prognosis [43]. Furthermore, Rakha *et al.* found that EGFR expression was associated with poor response to chemotherapy [44]. In addition, EGFR also plays a role in resistance to radiation treatment in TNBC [45].

5.3.3 *Ki67*

Ki67 is a protein that in humans is encoded by the MKI67 gene. Ki67 is highly associated with cell proliferation [46]. Ki67 can not only recognize cells in G_1, S, G_2 and M phases (except cells in the G_0 phase) to judge the proliferative activity of cells, but also be a marker for determining the growth state of benign and malignant tissues. In general, the higher level of Ki67 is associated with a higher level of malignancy and poorer prognosis, especially in certain types of cancer such as breast cancer and lymphoma [47, 48]. Ki67 is closely related to the histological classification, mitotic index, and lymph node metastasis of breast cancer. Ki67 is negatively correlated with the expression of hormone receptors. Therefore, it is an important reference indicator for judging tumor prognosis. It should be noted that Ki67 is not a good marker for subtyping of breast cancer since its expression can vary a lot from very little to ~80%.

Many studies have confirmed the potential use of Ki67 as a prognostic indicator and in predicting response to treatment in early breast cancer. However, Ki67 is not recommended for the management of early breast cancer patients due to variation in analytical practice. To address this problem, Syed *et al.* measured the Ki67 index in 119 patients and found that its significance correlates with known prognostic factors (such as ER, PR, and HER2 receptors) [49]. The results showed that the median value of Ki67 index was 20%, which was similar to other studies. However, the median Ki67 values were significantly increased in TNBC compared to other histologic types (70% in TNBC vs. 12.5% in luminal A, 20% in luminal B, and 30% in HER2-enriched subtype).

The optimal cutoff for Ki67 to predict TNBC outcomes was further assessed by scientists from Fudan University Shanghai Cancer Center using Cutoff Finder [50]. The most relevant cutoff value for Ki67 for prognosis was 30%. This cutoff value had early independent prognostic and predictive potential for OS and DFS in TNBCs. Ki67>30% was significantly associated with worse prognosis, especially for stage I TNBC patients. Therefore, it is important to measure the Ki67 index, which can be used as a marker in the treatment and follow-up of breast cancer.

5.3.4 *BRCA1/2*

The breast cancer susceptibility (BRCA) proteins, BRCA1 and BRCA2, play a critical role in the repair of DNA double-strand breaks through a conserved mechanism called homologous recombination (HR) [51]. While BRCA1 seems to have a relatively broader cellular functions besides DNA damage repair, such as transcriptional regulation and chromatin remodeling, BRCA2 function is largely restricted to DNA recombination and repair processes. Therefore, cells that lack functional BRCA1 or BRCA2 have a homologous recombination deficiency (HRD) in the repair of DNA (described more in Chapter 6). This deficiency results in the repair of DNA lesions by non-conservative and potentially mutagenic mechanisms such as non-homologous end joining and single strand annealing. HRD underlies the cancer predisposition caused by loss-of-function mutations of BRCA1 and BRCA2 [52].

Compared with the wild-type, BRCA mutations not only tend to be seen at younger patients, but also have a significantly better relapse-free survival (RFS) [53]. By analyzing BRCA1 and BRCA2 mutation from 199 patients with TNBC, Hartman *et al.* found about 4–5% risk of carrying a mutation if an individual was diagnosed with TNBC and did not have any family history of breast or ovarian cancer [54]. To determine the frequency of BRCA1 mutations among TNBC patients, Fostira *et al.* screened a large sample size of 403 women diagnosed with TNBC, independently of their age or family history, for germline BRCA1 mutations [55]. Thus, it is suggested that women diagnosed with TNBC younger than 50 years should be offered BRCA1 testing, regardless of family cancer characteristics. Furthermore, it was found that among women with early-onset breast cancer (≤35 years old) and with a family history of breast cancer, a higher prevalence of BRCA1 mutations was present in women with TNBC (42.1%) compared with women with non-TNBC (14.2%). The genetic testing of BRCA1/2 mutations is important for TNBC patients regardless of age and family history.

5.3.5 *PARPs*

Poly (ADP-ribose) polymerases (PARPs) are a family of proteins involved in a number of cellular processes such as DNA repair, genomic stability, and programmed cell death [56]. Among the PARP family proteins, PARP1 is a widely and abundantly expressed member. Some studies have shown that the activity of PARP is significantly enhanced in tumors, especially in tumors with BRCA mutation or loss of function [30, 57]. PARP1 plays a critical role in the initiation of DNA repair of single strand breaks by base excision repair pathway [58, 59]. When PARP1 is inhibited, further DNA damage can be induced as single-strand breaks, which can result in double-strand breaks during DNA replication. The DNA damage can be repaired through the HR mechanism (described above and more in Chapter 6) in the cells with normal BRCA functions [60]. However, since the BRCA1 and BRCA2 genes encode key components of the HR repair pathway, BRCA mutant tumors are inherently deficient in DNA repair, ultimately leading to cell death upon exposure to PARP inhibitors (PARPi's) [61]. Therefore, the PARPi's might selectively kill tumor cells while normal cells are not affected. This phenomenon, i.e., inactivation of individual genes does not lead to cell death but combining mutation or blockade of two or more genes leads to cell death, is called "synthetic lethality" (refer to Chapter 6 for more details) [59].

Some studies have shown that basal-like TNBC has high incidence of BRCA mutations and exhibits similar clinicopathological characteristics to BRCA1-mutated tumors [62–64]. Therefore, this high incidence of BRCA1 mutation provides a strategy for the treatment of TNBC with PARPi's. Hastak *et al.* found that TNBC cells were not only more sensitive to platinum and gemcitabine than non-TNBC cells, but also exhibited synergy with PARPi's in basal-like, but not in luminal, breast cancer cell lines [65]. Nowadays, PARPi's, such as olaparib, veliparib, and niraparib, have been selected for targeted treatment of TNBC [57–59, 66]. More in-depth and extensive research on PARP1 and its inhibitors will benefit a wider range of patients.

5.3.6 *Androgen Receptor*

Androgen receptor (AR), also known as NR3C4 (nuclear receptor subfamily 3 group C member 4), is a type of nuclear receptor that is activated by binding any of the androgenic hormones in the cytoplasm and then translocating into the nucleus. The main function of AR is as a DNA-binding transcription factor that regulates gene expression. AR and its ligand, androgen, are closely related to the development of breast cancer. The molecular mechanism of androgen on the development of breast cancer is mainly through the regulation of apoptotic gene family and growth factor signal transduction pathways. Androgen can regulate the expression of Bax and Bcl-2, unbalance the growth and apoptosis of cells, leading to rapid cell expansion.

On immunohistochemical analysis, approximately 10–15% of TNBC cases express AR [67, 68]. On the mRNA expression level, about 12% of TNBC cases express AR [28]. The subset of TNBC that expresses AR has been shown to express genes consistent with a luminal subtype and, therefore, has been subclassified as the luminal androgen receptor (LAR) subtype (discussed in Chapter 2).

The findings regarding the association between AR expression and prognostic value are inconsistent. AR-negative TNBC shows significantly poorer outcomes with regard to the DFS and OS than the AR-positive TNBC [44, 69]. Sutton *et al.* used immunohistochemistry to analyze AR in 121 TNBC patients and revealed that low expression of AR was associated with distant metastasis in the AR-positive TNBC cases [70]. Gasparini *et al.* analyzed AR expression by immunohistochemistry in 678 breast cancers and found that the expression level of AR was associated with better OS in the non-basal-like TNBC that had no expression of basal markers [71]. In addition, the association of AR status and breast cancer survival was dependent on ER expression. Among the women with ER-positive tumors, AR expression was associated with significantly improved survival [72]. However, the expression level of AR was lower in TNBC than non-TNBC [73]. Mrklić *et al.* found that there was no significant association between positive AR immunostaining and DFS or OS [74]. Recent investigations of the AR signaling pathway in breast cancer has rendered AR as a significant target for breast cancer therapy with several clinical trials currently in progress.

It should be noted that a subset of TNBC that lacks AR, which is termed quadruple-negative breast cancer (QNBC) by some researchers [75], predominantly exhibits a basal-like molecular subtype. Some important signaling molecules are being investigated as molecular targets in QNBC (discussed in Chapter 7).

5.3.7 *VEGF*

Vascular endothelial growth factor (VEGF), originally known as vascular permeability factor (VPF), is a signal protein produced by cells that stimulates the formation of blood vessels [76]. When VEGF binds to the receptor, it can stimulate the proliferation of vascular endothelial cells, promote blood vessel formation, and increase vascular permeability, so that tumor cells can not only gain sufficient nutrients to proliferate rapidly but also easily enter the blood through vascular endothelial cells to cause distant metastasis [77]. Breast cancer patients with high VEGF expression are prone to metastasis and recurrence following endocrine therapy and chemotherapy. Therefore, VEGF and its receptors can be used as important molecular markers for the treatment and prevention of breast cancer.

Compared with non-TNBC, TNBC patients have high VEGF expression and the high level of VEGF is associated with poor clinical outcome [78, 79]. In addition, VEGF is also significantly correlated with the size, grade, and metastasis of TNBC tumors [80]. Linderholm *et al.* analyzed 679 breast cancer patients and found that the median value of VEGF expression in the TNBC patients was 8.2 pg/μg DNA, while 2.7 pg/μg DNA in non-TNBC patients [81]. Bender *et al.* analyzed the expression of VEGF in more than 2,600 patients and found that TNBC was highly associated with dysregulation of VEGF-related genes. The expression of VEGF genes was altered in a pro-angiogenesis direction, and all TNBC groups demonstrated poor prognosis than non-TNBC samples [82]. In a study about the value of VEGF in patients with metastatic TNBC treated with fluorourcil, adriamycin, and cyclophamide (FAC) chemotherapy, Taha *et al.* found that VEGF level did not drop with the continuation of therapy and TNBC patients with high VEGF level had a significantly lower PFS but not OS than patients with low level. Furthermore, VEGF-targeting treatment is worth trying and therapies directed towards VEGF signaling axis to alleviate angiogenesis may be an alternative way to improve outcome for TNBC patients, which is discussed in Chapter 7.

5.3.8 *p53*

The p53 gene located on chromosome 17p13.1 encodes a nuclear phosphoprotein with a molecular weight of 53 kDa. Normally, p53 protein is activated during DNA damage or hypoxia, and p53-dependent cyclin-dependent kinase inhibitor p21 and DNA repair genes are up-regulated, and cells are arrested in the G_1 phase for DNA repair. If the repair is successful, the cells enter the S phase; if the repair fails, the cells are apoptotic by activating the Bax gene to ensure genetic stability of the genome. When the p53 gene is mutated or deleted, it will lose the negative regulatory function of mammary epithelial growth, leading to the occurrence and development of breast cancer.

At present, it is believed that the emergence of p53 mutant products is not only a specific biomarker of breast carcinogenesis, but also a very important reference biomarker for poor prognosis of breast cancer patients. Roos *et al.* indicated that p53 can be used as an independent prognostic biomarker for local recurrence of breast cancer at various stages, providing important reference for clinical treatment [83]. In a study by Chae *et al.*, p53 was strongly predictive for RFS and OS in TNBC patients, and p53 status could be a specific prognostic factor in TNBC patients treated by adjuvant anthracycline-based regimen [84]. However, Ko *et al.* indicated that expression of p53 was correlated with survival rate, patient's age, status of menopause, and tumor size, but it had no significance for the prognosis of breast cancer without axillary lymph node metastasis in the early stage [85]. In a study of 197 cases of TNBC patients in different races, Davion *et al.* found that p53 expression was more common in African Americans (77.7%) compared to Caucasians (57.1%), and TNBC patients younger than 50 (50%) were twice as likely to have p53 expression as those older than 50 (23.4%), suggesting that differences in p53 expression may be responsible for poor prognosis in African American TNBC patients [86]. Biganzoli *et al.* comparatively analyzed p53 expression from two independent breast cancer case series (total 1,709 patients) and found that TNBC patients with higher expression level of p53 exhibited worse overall and event-free survival compared to patients with lower p53 level [87].

5.3.9 Tyrosine Kinases

Tyrosine kinases (TKs) are a class of enzymes that can transfer a phosphate group from ATP to a protein in a cell. In the human genome, 90 TK genes and 5 presumed TK pseudogenes have been identified. These 90 TK genes include 58 receptor TKs and 32 non-receptor TKs (more details discussed in Chapter 7) [88]. They play important roles in the regulation of fundamental cellular processes such as cell development, differentiation, proliferation, survival, growth, apoptosis, etc [89]. Since their critical roles in normal cells, it is not surprising that when TKs are mutated, they will cause unregulated growth of the cell and cause cancer eventually. Clinical studies have shown that TK levels have a prognostic value in cancer patients [90].

By analyzing 26 TNBC cell lines, Wu *et al.* found that TK AXL (a member of the TAM receptor tyrosine kinase subfamily) is activated and highly expressed in most aggressive TNBC cell lines, and the high level of AXL is correlated with a significant decrease in patient survival, suggesting that inhibition of AXL has the potential to treat highly aggressive TNBC [91]. Hessel *et al.* analyzed 31 receptor TK-associated tumor relevant biomarkers in 29 TNBC patients using gene expression profiling and immunohistochemistry methods. They found that receptor TK expression is associated with survival and can be used for subdivision of TNBC into two subtypes where receptor TK high expression is associated with superior 3-year survival rate of 100% [92]. These studies suggested that TKs may have potential significance in new therapeutic approaches as an important novel biomarker [93, 94].

5.3.10 mTOR

Mammalian target of rapamycin (mTOR), also known as the mechanistic target of rapamycin and FK506-binding protein 12-rapamycin-associated protein 1 (FRAP1), is a kinase that in humans is encoded by the mTOR gene [95]. mTOR is an atypical serine/threonine kinase and belongs to the PI3K-related protein family involved in regulating major cellular functions including gene expression, cell growth and proliferation [96]. mTOR links with other proteins and serves as a core component of two distinct protein complexes, mTOR complex 1 (mTORC1) and mTOR complex 2 (mTORC2), which induces S-phase kinase association protein causing protein synthesis, proliferation, growth, metastasis, and angiogenesis [97, 98]. Deregulation of the mTOR signaling pathway is one of the most commonly observed pathological alterations in human cancers [99].

Since mTOR is an effector of PI3K signaling pathway regulated by Akt and the tumor-suppressor phosphatase and tensin homolog (PTEN), the proteins of the PI3K pathway are frequently affected by mutations in breast cancer and loss of PTEN is a common finding in TNBC, leading to the increased mTOR activation in TNBC [100]. Montero *et al.* observed co-activation of various receptor TKs in TNBC, together with frequent activation of PI3K/Akt/mTOR signaling pathway. Further pharmacologic studies showed that targeting mTOR signaling pathway is an effective treatment in TNBC than targeting receptor TK [101]. Furthermore, there is evidence that mTOR inhibitors such as everolimus (also known as RAD001) may have a role in TNBC. A phase II study showed that the addition of everolimus to standard chemotherapy in TNBC resulted in a small improvement in the 12-week response rate and was well tolerated by patients [102]. In addition, inhibition of the PI3K/mTOR pathway was identified as a promising therapeutic strategy for treating TNBC, and a phase II trial demonstrated that everolimus-carboplatin combination was efficacious in metastatic TNBC [103].

References

1. Mersin H, Yildirim E, Berberoglu U, Gülben K. (2008). The prognostic importance of triple negative breast carcinoma. *Breast*, 17(4): 341–346.
2. Kennecke H, Yerushalmi R, Woods R, Cheang MCU, Voduc D, Speers CH, Nielsen TO, Gelmon K. (2010). Metastatic behavior of breast cancer subtypes. *J Clin Oncol*, 28(20): 3271–3277.
3. O'Shaughnessy J, Osborne C, Pippen JE, Yoffe M, Patt D, Rocha C, Koo IC, Sherman BM, Bradley C. (2011). Iniparib plus chemotherapy

in metastatic triple-negative breast cancer. *New Engl J Med*, **364**(3): 205–214.

4. Rouzier R, Perou CM, Symmans WF, Ibrahim N, Cristofanilli M, Anderson K, Hess KR, Stec J, Ayers M, Wagner P *et al.* (2005). Breast cancer molecular subtypes respond differently to preoperative chemotherapy. *Clin Cancer Res*, **11**(16): 5678–5685.

5. Parker JS, Mullins M, Cheang MCU, Leung S, Voduc D, Vickery T, Davies S, Fauron C, He X, Hu Z *et al.* (2009). Supervised risk predictor of breast cancer based on intrinsic subtypes. *J Clin Oncol*, **27**(8): 1160–1167.

6. Kassam F, Enright K, Dent R, Dranitsaris G, Myers J, Flynn C, Fralick M, Kumar R, Clemons M. (2009). Survival outcomes for patients with metastatic triple-negative breast cancer: Implications for clinical practice and trial design. *Clin Breast Cancer*, **9**(1): 29–33.

7. Mattos-Arruda L, Cortes J, Santarpia L, Vivancos A, Tabernero J, Reis-Filho JS, Seoane J. (2013). Circulating tumour cells and cell-free DNA as tools for managing breast cancer. *Nat Rev Clin Oncol*, **10**: 377.

8. Mego M, Mani SA, Cristofanilli M. (2010). Molecular mechanisms of metastasis in breast cancer — clinical applications. *Nat Rev Clin Oncol*, **7**: 693.

9. Magbanua MJM, Carey LA, DeLuca A, Hwang J, Scott JH, Rimawi MF, Mayer EL, Marcom PK, Liu MC, Esteva FJ *et al.* (2015). Circulating tumor cell analysis in metastatic triple-negative breast cancers. *Clin Cancer Res*, **21**(5): 1098–1105.

10. Parkinson DR, Dracopoli N, Petty BG, Compton C, Cristofanilli M, Deisseroth A, Hayes DF, Kapke G, Kumar P, Lee JS *et al.* (2012). Considerations in the development of circulating tumor cell technology for clinical use. *J Transl Med*, **10**(1): 138.

11. Xenidis N, Ignatiadis M, Apostolaki S, Perraki M, Kalbakis K, Agelaki S, Stathopoulos EN, Chlouverakis G, Lianidou E, Kakolyris S *et al.* (2009). Cytokeratin-19 mRNA-positive circulating tumor cells after adjuvant chemotherapy in patients with early breast cancer. *J Clin Oncol*, **27**(13): 2177–2184.

12. Lianidou ES, Markou A. (2012). Molecular assays for the detection and characterization of CTCs. Sotiriou C, Pantel K. Berlin, Heidelberg: Springer Berlin Heidelberg; 2012: 111–123.

13. Cristofanilli M, Budd GT, Ellis MJ, Stopeck A, Matera J, Miller MC, Reuben JM, Doyle GV, Allard WJ, Terstappen LWMM *et al.* (2004).

Circulating tumor cells, disease progression, and survival in metastatic breast cancer. *New Engl J Med*, **351**(8): 781–791.

14. Hall C, Karhade M, Laubacher B, Anderson A, Kuerer H, DeSynder S, Lucci A. (2015). Circulating tumor cells after neoadjuvant chemotherapy in stage I–III triple-negative breast cancer. *Ann Surg Oncol*, **22 Suppl 3**: S552–558.

15. Lu Y-J, Wang P, Wang X, Peng J, Zhu Y-W, Shen N. (2016). The significant prognostic value of circulating tumor cells in triple-negative breast cancer: A meta-analysis. *Oncotarget*, **7**(24): 37361–37369.

16. Zhang Y, Lv Y, Niu Y, Su H, Feng A. (2017). Role of Circulating Tumor Cell (CTC) monitoring in evaluating prognosis of triple-negative breast cancer patients in China. *Med Sci Monit*, **23**: 3071–3079.

17. Schwarzenbach H, Hoon DSB, Pantel K. (2011). Cell-free nucleic acids as biomarkers in cancer patients. *Nat Rev Cancer*, **11**: 426.

18. Shaw JA, Stebbing J. (2013). Circulating free DNA in the management of breast cancer. *Ann Transl Med*, **2**(1): 3.

19. Diehl F, Schmidt K, Choti MA, Romans K, Goodman S, Li M, Thornton K, Agrawal N, Sokoll L, Szabo SA *et al.* (2008). Circulating mutant DNA to assess tumor dynamics. *Nat Med*, **14**: 985.

20. Bettegowda C, Sausen M, Leary RJ, Kinde I, Wang Y, Agrawal N, Bartlett BR, Wang H, Luber B, Alani RM *et al.* (2014). Detection of circulating tumor DNA in early- and late-stage human malignancies. *Sci Transl Med*, **6**(224): 224ra24.

21. Saliou A, Bidard F-C, Lantz O, Stern M-H, Vincent-Salomon A, Proudhon C, Pierga J-Y. (2016). Circulating tumor DNA for triple-negative breast cancer diagnosis and treatment decisions. *Expert Rev Mol Diagn*, **16**(1): 39–50.

22. Olsson E, Winter C, George A, Chen Y, Howlin J, Tang MHE, Dahlgren M, Schulz R, Grabau D, van Westen D *et al.* (2015). Serial monitoring of circulating tumor DNA in patients with primary breast cancer for detection of occult metastatic disease. *EMBO Mol Med*, **7**(8): 1034–1047.

23. Madic J, Kiialainen A, Bidard F-C, Birzele F, Ramey G, Leroy Q, Frio TR, Vaucher I, Raynal V, Bernard V *et al.* (2015). Circulating tumor DNA and circulating tumor cells in metastatic triple negative breast cancer patients. *Int J Cancer*, **136**(9): 2158–2165.

24. The Cancer Genome Atlas N, Koboldt DC, Fulton RS, McLellan MD, Schmidt H, Kalicki-Veizer J, McMichael JF, Fulton LL, Dooling DJ,

Ding L *et al.* (2012). Comprehensive molecular portraits of human breast tumours. *Nature*, **490**: 61.

25. Turner N, Lambros MB, Horlings HM, Pearson A, Sharpe R, Natrajan R, Geyer FC, van Kouwenhove M, Kreike B, Mackay A *et al.* (2010). Integrative molecular profiling of triple negative breast cancers identifies amplicon drivers and potential therapeutic targets. *Oncogene*, **29**: 2013.

26. Sharpe R, Pearson A, Herrera-Abreu MT, Johnson D, Mackay A, Welti JC, Natrajan R, Reynolds AR, Reis-Filho JS, Ashworth A *et al.* (2011). FGFR signaling promotes the growth of triple-negative and basal-like breast cancer cell lines both in vitro and in vivo. *Clin Cancer Res*, **17**(16): 5275–5286.

27. Bidard FC, Ng CKY, Cottu P, Piscuoglio S, Escalup L, Sakr RA, Reyal F, Mariani P, Lim R, Wang L *et al.* (2015). Response to dual HER2 blockade in a patient with HER3-mutant metastatic breast cancer. *Ann Oncol*, **26**(8): 1704–1709.

28. Lehmann BD, Bauer JA, Chen X, Sanders ME, Chakravarthy AB, Shyr Y, Pietenpol JA. (2011). Identification of human triple-negative breast cancer subtypes and preclinical models for selection of targeted therapies. *J Clin Invest*, **121**(7): 2750–2767.

29. Banerji S, Cibulskis K, Rangel-Escareno C, Brown KK, Carter SL, Frederick AM, Lawrence MS, Sivachenko AY, Sougnez C, Zou L *et al.* (2012). Sequence analysis of mutations and translocations across breast cancer subtypes. *Nature*, **486**: 405.

30. Fong PC, Boss DS, Yap TA, Tutt A, Wu P, Mergui-Roelvink M, Mortimer P, Swaisland H, Lau A, O'Connor MJ *et al.* (2009). Inhibition of poly(ADP-ribose) polymerase in tumors from BRCA mutation carriers. *New Engl J Med*, **361**(2): 123–134.

31. Tutt A, Robson M, Garber JE, Domchek SM, Audeh MW, Weitzel JN, Friedlander M, Arun B, Loman N, Schmutzler RK *et al.* (2010). Oral poly(ADP-ribose) polymerase inhibitor olaparib in patients with BRCA1 or BRCA2 mutations and advanced breast cancer: A proof-of-concept trial. *The Lancet*, **376**(9737): 235–244.

32. Schweizer J, Bowden PE, Coulombe PA, Langbein L, Lane EB, Magin TM, Maltais L, Omary MB, Parry DAD, Rogers MA *et al.* (2006). New consensus nomenclature for mammalian keratins. *J Cell Biol*, **174**(2): 169–174.

33. Araújo TG, Brandão DC, Marangoni K, Araújo GR, Maia YCP, Alves PT, Goulart LR. (2018). Transcripts of cytokeratins as predictors of breast cancer. *Gene Rep*, **13**: 14–18.

34. Otterbach F, Bànkfalvi À, Bergner S, Decker T, Krech R, Boecker W. (2000). Cytokeratin 5/6 immunohistochemistry assists the differential diagnosis of atypical proliferations of the breast. *Histopathology*, **37**(3): 232–240.

35. Ross DT, Perou CM. (2001). A comparison of gene expression signatures from breast tumors and breast tissue derived cell lines. *Dis Markers*, **17**(2): 99–109.

36. Abd El-Rehim DM, Pinder SE, Paish CE, Bell J, Blamey R, Robertson JF, Nicholson RI, Ellis IO. (2004). Expression of luminal and basal cytokeratins in human breast carcinoma. *J Pathol*, **203**(2): 661–671.

37. Sutton LM, Han JS, Molberg KH, Sarode VR, Cao D, Rakheja D, Sailors J, Peng Y. (2010). Intratumoral expression level of epidermal growth factor receptor and cytokeratin 5/6 is significantly associated with nodal and distant metastases in patients with basal-like triple-negative breast carcinoma. *Am J Clin Pathol*, **134**(5): 782–787.

38. Liu Z-B, Wu J, Ping B, Feng L-Q, Di G-H, Lu J-S, Shen K-W, Shen Z-Z, Shao Z-M. (2009). Basal cytokeratin expression in relation to immunohistochemical and clinical characterization in breast cancer patients with triple negative phenotype. *Tumori*, **95**(1): 53–62.

39. Thike AA, Iqbal J, Cheok PY, Chong APY, Tse GM-K, Tan B, Tan P, Wong NS, Tan PH. (2010). Triple negative breast cancer: Outcome correlation with immunohistochemical detection of basal markers. *Am J Surg Pathol*, **34**(7): 956–964.

40. Wells A. (1999). EGF receptor. *Int J Biochem Cell B*, **31**(6): 637–643.

41. Herbst RS, Langer CJ. (2002). Epidermal growth factor receptors as a target for cancer treatment: The emerging role of IMC-C225 in the treatment of lung and head and neck cancers. *Semin Oncol*, **29**(1, Supplement 4): 27–36.

42. Nogi H, Kobayashi T, Suzuki M, Tabei I, Kawase K, Toriumi Y, Fukushima H, Uchida K. (2009). EGFR as paradoxical predictor of chemosensitivity and outcome among triple-negative breast cancer. *Oncol Rep*, **21**(2): 413–417.

43. Viale G, Rotmensz N, Maisonneuve P, Bottiglieri L, Montagna E, Luini A, Veronesi P, Intra M, Torrisi R, Cardillo A *et al.* (2009). Invasive ductal carcinoma of the breast with the "triple-negative" phenotype: Prognostic implications of EGFR immunoreactivity. *Breast Cancer Res Tr*, **116**(2): 317–328.

44. Rakha EA, El-Sayed ME, Green AR, Lee AHS, Robertson JF, Ellis IO. (2007). Prognostic markers in triple-negative breast cancer. *Cancer*, **109**(1): 25–32.

45. Al-Ejeh F, Shi W, Miranda M, Simpson PT, Vargas AC, Song S, Wiegmans AP, Swarbrick A, Welm AL, Brown MP *et al.* (2013). Treatment of triple-negative breast cancer using anti-EGFR-directed radioimmunotherapy combined with radiosensitizing chemotherapy and PARP inhibitor. *J Nucl Med*, **54**(6): 913–921.

46. Scholzen T, Gerdes J. (2000). The Ki-67 protein: From the known and the unknown. *J Cell Physiol*, **182**(3): 311–322.

47. Azambuja E, Cardoso F, de Castro Jr G, Colozza M, Mano MS, Durbecq V, Sotiriou C, Larsimont D, Piccart-Gebhart MJ, Paesmans M. (2007). Ki-67 as prognostic marker in early breast cancer: A meta-analysis of published studies involving 12155 patients. *Brit J Cancer*, **96**: 1504.

48. Miglietta L, Morabito F, Provinciali N, Canobbio L, Meszaros P, Naso C, Murialdo R, Boitano M, Salvi S, Ferrarini M. (2013). A prognostic model based on combining estrogen receptor expression and Ki-67 value after neoadjuvant chemotherapy predicts clinical outcome in locally advanced breast cancer: Extension and analysis of a previously reported cohort of patients. *EJSO — Eur J Sur Onc*, **39**(10): 1046–1052.

49. Syed A, Giridhar PS, Sandhu K, Jader S, Al-Sam S, Sundaresan V, Singer J, Jenkins S, Bradpiece HA, Patel A. (2012). Ki67 in breast cancer patients and its correlation with clinico pathological factors. *Eur J Cancer*, **48**: S121.

50. Zhu X, Chen L, Huang B, Wang Y, Ji L, Wu J, Di G, Liu G, Yu K, Shao Z *et al.* (2020). The prognostic and predictive potential of Ki-67 in triple-negative breast cancer. *Sci Rep*, **10**(1): 225.

51. Turner N, Tutt A, Ashworth A. (2004). Hallmarks of 'BRCAness' in sporadic cancers. *Nat Rev Cancer*, **4**(10): 814–819.

52. Venkitaraman AR. (2002). Cancer susceptibility and the functions of BRCA1 and BRCA2. *Cell*, **108**(2): 171–182.

53. Gonzalez-Angulo AM, Timms KM, Liu S, Chen H, Litton JK, Potter J, Lanchbury JS, Stemke-Hale K, Hennessy BT, Arun BK *et al.* (2011). Incidence and outcome of BRCA mutations in unselected patients with triple receptor-negative breast cancer. *Clin Cancer Res*, **17**(5): 1082–1089.

54. Hartman A-R, Kaldate RR, Sailer LM, Painter L, Grier CE, Endsley RR, Griffin M, Hamilton SA, Frye CA, Silberman MA *et al.* (2012). Prevalence of BRCA mutations in an unselected population of triple-negative breast cancer. *Cancer*, **118**(11): 2787–2795.

55. Fostira F, Tsitlaidou M, Papadimitriou C, Pertesi M, Timotheadou E, Stavropoulou AV, Glentis S, Bournakis E, Bobos M, Pectasides D *et al.* (2012). Prevalence of BRCA1 mutations among 403 women with triple-negative breast cancer: Implications for genetic screening selection criteria: A hellenic cooperative oncology group study. *Breast Cancer Res Tr*, **134**(1): 353–362.

56. Herceg Z, Wang Z-Q. (2001). Functions of poly(ADP-ribose) polymerase (PARP) in DNA repair, genomic integrity and cell death. *Mutat Res*, **477**(1): 97–110.

57. Plummer R. (2011). Poly(ADP-ribose) polymerase inhibition: A new direction for BRCA and triple-negative breast cancer? *Breast Cancer Res*, **13**(4): 218.

58. Geenen JJJ, Linn SC, Beijnen JH, Schellens JHM. (2018). PARP inhibitors in the treatment of triple-negative breast cancer. *Clin Pharmacokinet*, **57**(4): 427–437.

59. Sonnenblick A, de Azambuja E, Azim Jr HA, Piccart M. (2014). An update on PARP inhibitors — moving to the adjuvant setting. *Nat Rev Clin Oncol*, **12**: 27.

60. Shall S, de Murcia G. (2000). Poly(ADP-ribose) polymerase-1: What have we learned from the deficient mouse model? *Mutat Res*, **460**(1): 1–15.

61. Bryant HE, Schultz N, Thomas HD, Parker KM, Flower D, Lopez E, Kyle S, Meuth M, Curtin NJ, Helleday T. (2005). Specific killing of BRCA2-deficient tumours with inhibitors of poly(ADP-ribose) polymerase. *Nature*, **434**: 913.

62. Holstege H, Horlings HM, Velds A, Langerød A, Børresen-Dale A-L, van de Vijver MJ, Nederlof PM, Jonkers J. (2010). BRCA1-mutated and basal-like breast cancers have similar aCGH profiles and a high incidence of protein truncating TP53 mutations. *BMC Cancer*, **10**(1): 654.

63. Hill SJ, Clark AP, Silver DP, Livingston DM. (2014). BRCA1 pathway function in basal-like breast cancer cells. *Mol Cell Biol*, **34**(20): 3828–3842.

64. Turner NC, Reis-Filho JS. (2006). Basal-like breast cancer and the BRCA1 phenotype. *Oncogene*, **25**: 5846.

65. Hastak K, Alli E, Ford JM. (2010). Synergistic chemosensitivity of triple-negative breast cancer cell lines to poly(ADP-Ribose) polymerase inhibition, gemcitabine, and cisplatin. *Cancer Res*, **70**(20): 7970–7980.

66. Rouleau M, Patel A, Hendzel MJ, Kaufmann SH, Poirier GG. (2010). PARP inhibition: PARP1 and beyond. *Nat Rev Cancer*, **10**: 293.

67. Gucalp A, Traina TA. (2010). Triple-negative breast cancer: Role of the androgen receptor. *Cancer J*, **16**(1): 62–65.

68. Doane AS, Danso M, Lal P, Donaton M, Zhang L, Hudis C, Gerald WL. (2006). An estrogen receptor-negative breast cancer subset characterized by a hormonally regulated transcriptional program and response to androgen. *Oncogene*, **25**: 3994.

69. Tang D, Xu S, Zhang Q, Zhao W. (2012). The expression and clinical significance of the androgen receptor and E-cadherin in triple-negative breast cancer. *Med Oncol*, **29**(2): 526–533.

70. Sutton LM, Cao D, Sarode V, Molberg KH, Torgbe K, Haley B, Peng Y. (2012). Decreased androgen receptor expression is associated with distant metastases in patients with androgen receptor-expressing triple-negative breast carcinoma. *Am J Clin Pathol*, **138**(4): 511–516.

71. Gasparini P, Fassan M, Cascione L, Guler G, Balci S, Irkkan C, Paisie C, Lovat F, Morrison C, Zhang J *et al.* (2014). Androgen receptor status is a prognostic marker in non-basal triple negative breast cancers and determines novel therapeutic options. *PloS one*, **9**(2): e88525.

72. Hu R, Dawood S, Holmes MD, Collins LC, Schnitt SJ, Cole K, Marotti JD, Hankinson SE, Colditz GA, Tamimi RM. (2011). Androgen receptor expression and breast cancer survival in postmenopausal women. *Clin Cancer Res*, **17**(7): 1867–1874.

73. Zhang L, Fang C, Xu X, Li A, Cai Q, Long X. (2015). Androgen receptor, EGFR, and BRCA1 as biomarkers in triple-negative breast cancer: A meta-analysis. *Biomed Res Int*, **2015**: 357485.

74. Mrklić I, Pogorelić Z, Ćapkun V, Tomić S. (2013). Expression of androgen receptors in triple negative breast carcinomas. *Acta Histochem*, **115**(4): 344–348.

75. Hon JD, Singh B, Sahin A, Du G, Wang J, Wang VY, Deng FM, Zhang DY, Monaco ME, Lee P. (2016). Breast cancer molecular subtypes: From TNBC to QNBC. *Am J Cancer Res*, **6**(9): 1864–1872.

76. Senger DR, Galli SJ, Dvorak AM, Perruzzi CA, Harvey VS, Dvorak HF. (1983). Tumor cells secrete a vascular permeability factor that promotes accumulation of ascites fluid. *Science*, **219**(4587): 983–985.

77. El-Arab LRE, Swellam M, El Mahdy MM. (2012). Metronomic chemotherapy in metastatic breast cancer: Impact on VEGF. *J Egypt Natl Canc Inst*, **24**(1): 15–22.

78. Taha FM, Zeeneldin AA, Helal AM, Gaber AA, Sallam YA, Ramadan H, Moneer MM. (2009). Prognostic value of serum vascular endothelial growth factor in Egyptian females with metastatic triple negative breast cancer. *Clin Biochem*, **42**(13): 1420–1426.

79. Dent SF. (2009). The role of VEGF in triple-negative breast cancer: Where do we go from here? *Ann Oncol*, **20**(10): 1615–1617.

80. Chanana P, Pandey AK, Yadav BS, Kaur J, Singla S, Dimri K, Trehan R, Krishan P. (2013). Significance of serum vascular endothelial growth factor and cancer antigen 15.3 in patients with triple negative breast cancer. *Journal of Radiotherapy in Practice*, **13**(1): 60–67.

81. Linderholm BK, Hellborg H, Johansson U, Elmberger G, Skoog L, Lehtio J, Lewensohn R. (2009). Significantly higher levels of vascular endothelial growth factor (VEGF) and shorter survival times for patients with primary operable triple-negative breast cancer. *Ann Oncol*, **20**(10): 1639–1646.

82. Bender RJ, Mac Gabhann F. (2013). Expression of VEGF and sema-phorin genes define subgroups of triple negative breast cancer. *PloS One*, **8**(5): e61788.

83. Roos MA, de Bock GH, de Vries J, van der Vegt B, Wesseling J. (2007). P53 overexpression is a predictor of local recurrence after treatment for both in situ and invasive ductal carcinoma of the breast. *J Surg Res*, **140**(1): 109–114.

84. Chae BJ, Bae JS, Lee A, Park WC, Seo YJ, Song BJ, Kim JS, Jung SS. (2009). P53 as a specific prognostic factor in triple-negative breast cancer. *Jpn J Clin Oncol*, **39**(4): 217–224.

85. Ko S-S, Na Y-S, Yoon C-S, Park J-Y, Kim H-S, Hur M-H, Lee H-K, Chun Y-K, Kang S-S, Park B-W *et al.* (2007). The significance of c-erbB-2 overexpression and p53 expression in patients with axillary lymph node — negative breast cancer: A tissue microarray study. *Int J Surg Pathol*, **15**(2): 98–109.

86. Davion S, Sullivan M, Rohan S, Siziopikou KP. (2012). P53 expression in triple negative breast carcinomas: Evidence of age-related and racial differences. *Journal of Cancer Therapy*, **3**(5A): 649–654.

87. Biganzoli E, Coradini D, Ambrogi F, Garibaldi JM, Lisboa P, Soria D, Green AR, Pedriali M, Piantelli M, Querzoli P *et al.* (2011). p53 status identifies two subgroups of triple-negative breast cancers with distinct biological features. *Jpn J Clin Oncol*, **41**(2): 172–179.

88. Robinson DR, Wu Y-M, Lin S-F. (2000). The protein tyrosine kinase family of the human genome. *Oncogene*, **19**: 5548.

89. Hunter T. (1998). The croonian lecture 1997. The phosphorylation of proteins on tyrosine: Its role in cell growth and disease. *Philos Trans R Soc Lond B Biol Sci*, **353**(1368): 583–605.

90. Madhusudan S, Ganesan TS. (2004). Tyrosine kinase inhibitors in cancer therapy. *Clin Biochem*, **37**(7): 618–635.

91. Wu X, Zahari MS, Ma B, Liu R, Renuse S, Sahasrabuddhe NA, Chen L, Chaerkady R, Kim M-S, Zhong J *et al.* (2015). Global phosphotyrosine survey in triple-negative breast cancer reveals activation of multiple tyrosine kinase signaling pathways. *Oncotarget*, **6**(30): 29143–29160.

92. Hessel H, Poignée-Heger M, Lohmann S, Hirscher B, Herold A, Assmann G, Budczies J, Sotlar K, Kirchner T. (2018). Subtyping of triple negative breast carcinoma on the basis of RTK expression. *J Cancer*, **9**: 2589–2602.

93. Ho-Yen CM, Jones JL, Kermorgant S. (2015). The clinical and functional significance of c-Met in breast cancer: A review. *Breast Cancer Res*, **17**(1): 52.

94. Nunes-Xavier CE, Martín-Pérez J, Elson A, Pulido R. (2013). Protein tyrosine phosphatases as novel targets in breast cancer therapy. *BBA — Rev Cancer*, **1836**(2): 211–226.

95. Moore PA, Rosen CA, Carter KC. (1996). Assignment of the human FKBP12-rapamycin-associated protein (FRAP) gene to chromosome 1p36 by fluorescence in situ hybridization. *Genomics*, **33**(2): 331–332.

96. Hay N, Sonenberg N. (2004). Upstream and downstream of mTOR. *Genes Dev*, **18**(16): 1926–1945.

97. Lipton Jonathan O, Sahin M. (2014). The neurology of mTOR. *Neuron*, **84**(2): 275–291.

98. Wullschleger S, Loewith R, Hall MN. (2006). TOR signaling in growth and metabolism. *Cell*, **124**(3): 471–484.

99. Xu K, Liu P, Wei W. (2014). mTOR signaling in tumorigenesis. *BBA — Rev Cancer*, **1846**(2): 638–654.

100. Saal LH, Holm K, Maurer M, Memeo L, Su T, Wang X, Yu JS, Malmström P-O, Mansukhani M, Enoksson J *et al.* (2005). PIK3CA mutations correlate with hormone receptors, node metastasis, and ERBB2, and are mutually exclusive with PTEN loss in human breast carcinoma. *Cancer Res*, **65**(7): 2554–2559.

101. Montero JC, Esparís-Ogando A, Re-Louhau MF, Seoane S, Abad M, Calero R, Ocaña A, Pandiella A. (2012). Active kinase profiling, genetic and pharmacological data define mTOR as an important common target in triple-negative breast cancer. *Oncogene*, **33**: 148.

102. Gonzalez-Angulo AM, Green MC, Murray JL, Palla SL, Koenig KH, Brewster AM, Valero V, Ibrahim NK, Moulder SL, Litton JK *et al.* (2011). Open label, randomized clinical trial of standard neoadjuvant chemotherapy with paclitaxel followed by FEC (T-FEC) versus the combination of paclitaxel and RAD001 followed by FEC (TR-FEC) in women with triple receptor-negative breast cancer (TNBC). *J Clin Oncol*, **29**(15_suppl): 1016–1016.

103. Singh JC, Novik Y, Stein S, Volm M, Meyers M, Smith J, Omene C, Speyer J, Schneider R, Jhaveri K *et al.* (2014). Phase 2 trial of everolimus and carboplatin combination in patients with triple negative metastatic breast cancer. *Breast Cancer Res*, **16**(2): R32.

Chapter SIX

Clinical Aspects of Triple-Negative Breast Cancer

Doudou Huang[1], Jingyang Du[2,3],
Xiaoxiang Guan[4,*], *and* Xiyun Deng[2,3,*]

Contents

*Corresponding authors: Xiaoxiang Guan, E-mail: xguan@nju.edu.cn; Xiyun Deng, E-mail: dengxiyunmed@hunnu.edu.cn

[1]Department of Medical Oncology, Jinling Hospital, Medical School of Nanjing University, Nanjing, China.
[2]Key Laboratory of Translational Cancer Stem Cell Research, Hunan Normal University, Changsha, Hunan, China.
[3]Departments of Pathology and Pathophysiology, Hunan Normal University School of Medicine, Changsha, Hunan, China.
[4]Department of Oncology, The First Affiliated Hospital of Nanjing Medical University, Nanjing, China.

Defined as a subtype which is negative for ER, PR, and HER2, triple-negative breast cancer (TNBC) demonstrates a more aggressive clinical behavior and a poorer outcome [1]. Responsible for lots of deaths from breast cancer, TNBC predicts a short disease-free survival (DFS) and overall survival (OS). In addition, TNBC has some unique pathological characteristics with a high heterogeneity, which has pushed plenty of studies to be carried out to explore the novel treatment modalities of this subtype. A comprehensive understanding of the heterogeneous nature of TNBC would assuredly contribute to the precision diagnosis and individualized therapy of this difficult-to-treat subtype. Some of the features and diagnosis criteria of TNBC/BLBC have been described in Chapter 1. This chapter focuses on more clinical aspects of TNBC, regarding its clinicopathological features, diagnosis, treatment, prognosis, and nursing implications.

6.1 Clinicopathological Features of TNBC

6.1.1 *General Clinical Features*

Generally speaking, TNBC is associated with younger age and higher probability of relapse and is more prevalent in African American ethnicity and BRCA1/2 mutation carriers [2]. The risk of early relapse is substantially high, with the peak recurrence between the first and the third year after diagnosis, followed by a decrease of the recurrence rate [3]. Compared with ER-positive breast cancer, TNBC experiences a worse survival and a high death rate in the first five years after treatment. However, the survival rate approaches that of other breast cancer subtypes after five years [4].

In terms of its highly malignant and aggressive clinical behaviors, rapidly developed drug resistance and metastasis are commonly observed in patients with TNBC. Distant metastases are prone to develop before locoregional recurrence. As will be discussed under Section 6.1.3, of all the organs, lungs and brain are more likely to be affected than bone and liver which are the commonly metastasized sites for non-TNBC [5]. The risk of distant metastasis increases rapidly in the first two years after diagnosis, with the highest between the second and third year, then declining after two more years. In addition, following distant metastasis, the median duration of survival is around ½ for basal-like breast cancers (BLBCs) [6, 7]. Different from other subtypes, however, recurrence rate decreases sharply after eight years [3, 7–9].

Multiple studies have demonstrated that TNBC disproportionately affects women of African ancestry, though there regionally exists a variation of TNBC incidence even in the same ethnic group. According to the US Surveillance, Epidemiology, and End Results (SEER) California Cancer Registry database, the incidence of TNBC vs. luminal subtype among African-American women was 1:2, while the ratio was significantly lower in other ethnic groups (1:6.9 and 1:6.1 in white and Asian women, respectively) [10]. Such high incidence of TNBC in women of African ancestry has also been found in large population-based studies [11]. Of note, there is evidence showing that in sub-Sahara and West African women, TNBC is more frequently observed than in African-American women, as well as the younger age at diagnosis. In spite of the comparably low incidence of breast cancer among women of sub-Sahara Africa (10–40 in 100,000), the mortality rate (5–20 in 100,000) is high [11].

The BRCA1/2 genes are important breast cancer susceptibility genes (more discussed in Chapter 5 and Chapter 7). It is reported that

Table 6-1.   General clinical features of TNBC.

Younger age at presentation
Aggressive behavior
Overlap with BLBC
Association with BRCA1/2 gene defects
Shorter time to relapse
Higher risk of visceral metastasis (including lung, liver)
Drug resistance
Worse prognosis

Refs: *Nat Rev Clin Oncol. 2010; 7(12):683–692; Pharmacol Ther. 2017; 175:91–106; Oncologist. 2016; 21(9):1050–1062.*

nearly 68% patients of breast cancer with BRCA1 mutation fall into the category of TNBC, while the percentage for BRCA2 mutation is 16% [12]. Additionally, distribution of BRCA1 mutation has an ethnic disparity, with the incidence of germline mutation being lower in women of African origin than those of European origin [13, 14]. The general clinical features of TNBC are summarized in **Table 6-1**.

6.1.2 *Pathological Features and Immunophenotypes*

Compared with hormone receptor-positive patients, TNBC patients tend to be of high grade, have lymphovascular invasion, and present with clinically metastatic disease. Histologically, the majority of TNBC patients present with an invasive ductal feature of no special type, but typical or atypical medullary, metaplastic and adenoid cystic elements are occasionally observed (discussed in Chapter 2). Additionally, basal-like phenotype is also found in most of metaplastic and medullary carcinomas. With regard to morphological features, central necrosis, a pushing border of invasion and pronounced lymphocytic infiltration, as well as apoptotic cells, are also characteristics of TNBC, together with high histologic grade (mostly grade 3, some grade 2), high mitotic index, and expression of EMT markers [15]. In addition, TNBC also has the characteristic of high degree of aneuploidy and nuclear pleomorphism. The pathological characteristics of TNBC/BLBC are summarized in **Table 6-2**.

Table 6-2. Pathological characteristics of TNBC/BLBC.

High grade

Lymphovascular invasion

Clinical metastasis

Mostly belonging to IDC

Elevated mitotic count/High proliferation rate

Central tumor necrosis

Pushing borders of invasion

Dense lymphocytic infiltration

High degree of aneuploidy

Nuclear pleomorphism

Refs: *Clin Breast Cancer. 2009; 9(Suppl 2):S73–S81; Semin Oncol. 2011; 38(2):254–262.*

As its name indicates, TNBC is absent of ER, PR, and HER2 expression revealed by immunohistochemistry and/or fluorescence in situ hybridization (FISH) [16]. Because of the overlaps between TNBC and BLBC (discussed in Chapter 1), their immunophenotypes also show a very similar pattern. The staining for EGFR, cytokeratins (CKs) 5, 6, 17 and c-Kit are often positive in both TNBC and BLBC. In addition, protein abnormality or gene mutation of p53 occurs in more than half of TNBC tumors. Though the activating mutation rate of PIK3CA is just about 10% in primary TNBC, the PI3K pathway is commonly activated in TNBC. The immunophenotypic and molecular characteristics of TNBC/BLBC are summarized in **Table 6-3**.

6.1.3 *Increased Risk of Pulmonary and Brain Metastasis in TNBC*

Highly invasive, TNBC develops a distinct locoregional and distant metastasis pattern, commonly spreading to the lungs, the brain, and the liver. The involvement of bone is relatively less found especially compared with non-TNBC [5] (**Figure 6-1**). While pulmonary metastasis is usually not found in luminal A subtype, its presence in

Table 6-3. Immunophenotypic and molecular characteristics of TNBC.

ER negative

PR negative

HER2 negative

EGFR and/or CK5/6 positive

c-Kit positive

Activation of the PI3K pathway

High expression of stemness and EMT markers

Refs: *Clin Breast Cancer. 2009; 9(Suppl 2):S73–S81; Oncologist. 2016; 21(9): 1050–1062.*

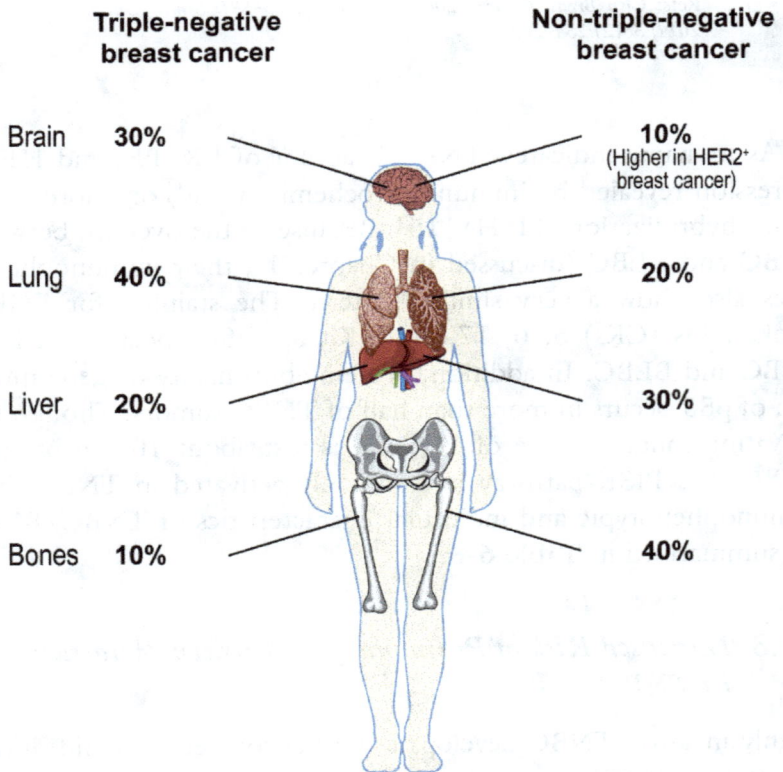

	Triple-negative breast cancer	Non-triple-negative breast cancer
Brain	30%	10% (Higher in HER2⁺ breast cancer)
Lung	40%	20%
Liver	20%	30%
Bones	10%	40%

Figure 6-1. Common sites of metastasis of TNBC vs. non-TNBC. The percentages shown are approximate percentages of breast cancer patients with a first distant recurrence among women in whom metastases develop.

TNBC is comparatively typical. There are several characteristics for pulmonary metastasis in medical imaging: solitary or multiple pulmonary nodules, lymphangitic carcinomatosis, endobronchial metastases, and air-space consolidation. Apart from these, pleural involvement, mostly manifesting as a pleural effusion, is frequently unilateral and ipsilateral to the breast cancer. Nevertheless, there is no specific characteristic to distinguish this pleural disease from benign effusions arising from other conditions on either X-ray or CT scan [17].

As for brain metastasis, it is reported that in TNBC the risk is approximately 6–46% of all the metastases and is conspicuous in younger patients [15]. In clinical settings, given the higher resolution of soft tissue and more accurate delineation of parenchymal and leptomeningeal, magnetic resonance imaging (MRI) is better than CT in identifying brain metastasis of TNBC. Accompanied with manifestations such as headache and mental status changes, brain lesions are often observed in the cerebral hemispheres as solitary or multiple metastases. In addition, such lesions are inclined to occurring at the grey-white junction and watershed areas of the major arterial domains [18].

Unlike metastasis in HER2-positive breast cancer, central nervous spread arising from TNBC indicates worse or progressive condition [19]. Thus, what the diagnosis of intracranial metastasis follows is mostly a shorter median survival of 3–5 months in TNBC vs. 7–12 months in non-TNBC [20]. It should be noted that although a significant portion of breast cancers metastasize to the lymph node, TNBC does not have increased likelihood of lymph node metastases compared with other subtypes of breast cancer [21].

6.2 Diagnosis of TNBC

6.2.1 *Diagnosis Based on the Status of Receptors*

To make a diagnosis of TNBC, the status of the three receptors, i.e., ER, PR, and HER2, are tested. As mentioned in Chapter 1, since the gene expression microarray analysis is costly and complicated, in the clinical setting, immunohistochemistry with or without FISH is usually performed instead to reveal the status of these receptors. It is worth noting that when immunohistochemical analysis is conducted, the cutoffs for ER, PR, and HER2 status were applied to postulate the likelihood of response to endocrine and anti-HER2 therapies, respectively, but not to determine the "absolute" phenotype. Therefore, the cutoffs that were developed to make the diagnosis have a significant impact on determining the status of the receptors, and therefore, the number of TNBC cases diagnosed. For now, the standard procedure most studies have adopted is based on the current guidelines recommended by the American Society of Clinical Oncology (ASCO) and College of American Pathologists (CAP).

To determine ER/PR negativity, ER and PR nuclear staining was initially set at <10% in terms of positively stained cells. It was later found that breast tumors with very low levels of ER/PR (between 1% and 10% positive cells) show a significant benefit from hormonal therapies. Therefore, the ASCO/CAP recommended adopting <1% as the definition of ER/PR-negative staining. The antibodies recommended for ER testing include ID5, 6F11, SP1, or ER.2.123 + 1D5 (cocktail) and the antibodies for PR testing include 1294, 1A6, or 312 [22].

With regard to HER2 testing, the widely accepted guideline is the 2013 updates to the ASCO/CAP recommendations for HER2 testing in breast cancer [23, 24]. According to these recommendations, HER2-positive status is defined when (on observing within an area of tumor that amounts to >10% of contiguous and homogeneous tumor cells) there is evidence of protein overexpression determined by immunohistochemistry or gene amplification (HER2 copy number) determined by FISH. For the samples with equivocal HER2 status in immunohistochemistry, further analysis by FISH should be performed in order to confirm the HER2 status and vice versa. Furthermore, repeat testing should be considered if the results seem discordant with other histopathological findings.

6.2.2 *Medical Imaging as Complementary Diagnostic Methods*

Additionally, imaging also contributes to complementary diagnosis of TNBC. Magnetic resonance imaging (MRI) is reported to be very sensitive for detection of tumor mass [25]. TNBC presents specific characteristics on MRI, such as smooth margin, rim enhancement, and very high intratumoral signal intensity on T2-weighted MRIs [15–17]. Dynamic contrast-enhanced MRI (DCE-MRI), by comparing the MRI features before and after neoadjuvant chemotherapy, could predict the pathologic complete response (pCR) of tumors to chemotherapy [26, 27]. Ultrasound is another way to make diagnostic evaluation. In ultrasound, the most characteristic features of TNBC are lesions with well-circumscribed margin and posterior acoustic enhancement [25].

Shear wave elastography and contrast-enhanced ultrasound are also under evaluation as tools for discriminating TNBC from other subtypes of breast cancer [28, 29]. In addition to MRI and ultrasound, mammography is commonly used in breast cancer diagnosis. In mammography, the TNBC mass is mostly observed in round or oval shape.

6.3 Treatment of TNBC

Although several recommendations have been proposed for the treatment of TNBC, currently, there is no widely accepted guideline for TNBC patients [30]. Due to the lack of ER, PR, and HER2, TNBC does not respond to hormonal and HER2-targeted therapies which are effective in non-TNBC. Treatment modalities for TNBC consist of two parts, locoregional treatment (including surgery and radiotherapy) and systemic chemotherapy. Hence, combined with surgery, chemotherapy lays the foundations for TNBC treatment. Despite initial high response rates, unfortunately, relapse rates are high in patients who do not achieve a pCR, resulting in a worse OS in patients with TNBC and BLBC compared with other subtypes of breast cancer. This chapter mainly focuses on the currently available treatment modalities of TNBC. Novel therapeutic options under investigation will be discussed in Chapter 7 (focusing on targeted therapies against signaling pathways and cancer stem cells) and Chapter 8 (focusing on immune checkpoint inhibition).

6.3.1 *Surgery*

Since TNBC is much more aggressive than other subtypes, many studies have been conducted to determine whether mastectomy would be superior over breast conservative surgery (BCS) such as lumpectomy, in which only the discrete portion is removed. However, it has been demonstrated that BCS provides at least equivalent prognosis to mastectomy, i.e., removal of the full breast. As a result, instead of TN status, final decision about the surgical method would depend more on traditional clinicopathological variables and patients' preference [31]. What is more, though highly malignant and with poor OS, it is found that, in TNBC, the local recurrence rate is not significantly higher than those of other subtypes after BCS [32]. Moreover, using the SEER database enrolling 11,514 TNBC cases, Chen *et al.* found that patients with BCS plus radiotherapy exhibited better breast cancer-specific survival and OS than patients who received mastectomy [33]. Therefore, when it comes to patients with TNBC, BCS might also be taken into consideration [32].

6.3.2 *Chemotherapy*

Compared with locoregional treatments such as surgery and radiotherapy, systemic treatment is directed against genetic aberrations and the molecular status of the tumor. In general, TNBC is more responsive to chemotherapy than any other subtype of breast cancer. Patients with breast cancer including TNBC who achieve pCR after neoadjuvant treatment have improved long-term outcomes. Even so, TNBC is still prone to metastasis and recurrence, especially in those who do not achieve pCR. This phenomenon of inconsistency between chemotherapy response and clinical outcome has been called a "triple-negative paradox" [34].

Chemotherapeutic agents for the management of TNBC act by specific mechanisms: microtubule-interfering (e.g., taxanes), cell proliferation inhibition (e.g., anthracycline-based regimens), and DNA damage (e.g., platinum compounds) [31, 35]. Taxanes (paclitaxel and docetaxel), which function by interfering mitotic spindle component, are widely used first-line chemotherapeutic drugs for TNBC. Accumulated evidence has demonstrated the efficacy of taxanes in TNBC rather than non-TNBC. Anthracyclines (doxorubicin and epirubicin) are considered to be among the most active drugs for the treatment of breast cancer. Many studies have shown that TNBC is sensitive to anthracycline-containing regimens [36]. Platinum-based chemotherapy is a commonly used second-line regimen. As DNA damaging agents, platinum compounds can cause cell death via interaction with DNA. Emerging data has confirmed the association between BRCA1 mutation, which could result in DNA repair dysfunction, with increased sensitivity to DNA damaging therapeutics, spurring a wave of new exploration in platinum agents both in preclinical and clinical studies [31]. Etoposide and bleomycin could add sensitivity to platinum-based regimen by inducing DNA double-strand breaks.

In TNBC patients without BRCA mutation, single-agent chemotherapy with taxanes has been recommended as a first-line treatment. In patients with a high disease burden or who are very symptomatic, combinations such as anthracyclines plus cyclophosphamide or platinum with taxanes are thought to be valid options. For patients who develop resistance or present contraindications to first-line treatments, fluorouracil/capecitabine, eribulin, gemcitabine, cisplatin/carboplatin, vinorelbine and ixabepilone are considered alternatives [37].

6.3.2.1 *Neoadjuvant chemotherapy*

Aiming at reducing tumor size or extent to render an inoperable tumor operable and even allow more conservative surgery, neoadjuvant chemotherapy is a new strategy introduced in the late 20th century and is being increasingly performed [38]. Unlike traditional adjuvant chemotherapy, neoadjuvant chemotherapy is based on smaller patient accruals and a shorter period to make a rapid assessment of treatment efficacy by determining in vivo tumor response and pCR. It has been shown that neoadjuvant chemotherapy could increase response rates in TNBC, and pCR might be regarded as a potential predictor indicating a better long-term outcome [39].

Multiple studies have demonstrated that application of platinum could improve pCR, rendering platinum-based regimens as a promising treatment for TNBC. Because of the genetic instability such as BRCA1 mutation, TNBC patients might be particularly sensitive to platinum agents [31, 40]. In addition, the efficacy of neoadjuvant chemotherapy including anthracyclines and taxanes has also been confirmed in TNBC with substantially higher pCR rates observed in comparison with non-TNBC [41]. However, despite the rapid response brought by neoadjuvant chemotherapy, no evidence of survival benefit has been recognized but a higher rate of local recurrence because of the more conservative surgery, which indicates that pCR is not a reliable surrogate maker for chemotherapy regimen selection. More seriously, along with the tumor shrunk by the therapeutics is a more difficult surgery [31, 42].

6.3.2.2 *Adjuvant chemotherapy*

Currently, multiple studies have confirmed the long-term benefit conferred by adjuvant chemotherapy for patients with TNBC. The regimen for moderate-to-high risk TNBC is a sequential anthracycline-cyclophosphamide-taxane (ACT) combination. In patients with early-stage or low-risk TNBC, cyclophosphamide-methotrexate-fluorouracil (CMF) is the traditional protocol. Nevertheless, it was observed that anthracycline-cyclophosphamide (AC) could be similar to the CMF regimen, in terms of efficiency and side effects.

In addition, consistent with findings in the neoadjuvant setting, mutant BRCA1 carriers with TNBC also might benefit from platinum compounds [31, 43]. Interestingly, in vitro studies showed that taxanes, unlike platinum compounds, seem to be less effective in treating breast cancer with BRCA1 mutations [31]. Nevertheless, currently, there are no convincing clinical data that confirm a potential resistance to taxanes in TNBC. In fact, recent studies have demonstrated the advantage of using taxanes in adjuvant setting [31].

As for anthracyclines, which have been applied to breast cancer treatment for several decades particularly because of their substantial impact in breast cancer overexpressing HER2, whether there is a superior efficacy in treating TNBC remains controversial [31]. Intriguingly, it is argued that topoisomerase IIa enzyme might be a potential predictor for response to anthracyclines. However, anthracyclines are found to induce the alterations of topoisomerase IIa enzyme, which might eventually drive the anthracycline resistance [44]. In addition, for BRCA1-mutated TNBC, whether or not it could gain more benefit from anthracyclines still remains unclear [31].

In addition to the chemotherapeutic agents mentioned above, with evidence from subgroup analysis, capecitabine, one of antimetabolites, might also show activity as an extra addition to regimens with anthracyclines and taxanes. In contrast, there are also data showing an inferior effect with capecitabine. Due to the disparity, more studies should be carried out to confirm the efficacy of capecitabine in the adjuvant setting [31].

It should be noted that although numerous large-scale randomized trials have been performed and established, the benefit of adjuvant chemotherapy for TNBC, several issues still remain unsolved [30]. First, there is still no widely accepted optimal or preferred regimen using chemotherapeutic agents. Second, it is not known whether adding agents such as platinum will benefit TNBC patients. Third, the best duration of adjuvant therapy in TNBC has yet to be determined.

6.3.2.3 *Chemoresistance in TNBC*

Along with the prolonged survival following chemotherapy, the resistance to chemotherapeutic agents occurs, which accounts for 90% of treatment failures in the metastatic setting. By undergoing alternate mechanisms to escape from cytotoxic chemotherapy, cancer cells are able to maintain viability. Therefore, a clear comprehension of the underlying mechanisms is needed so as to shed light on development of corresponding therapies to overcome the resistance and provide better treatment for patients with TNBC.

Mechanisms of chemoresistance

Currently, a variety of mechanisms have been proposed to confer resistance to chemotherapy in TNBC [45].

(1) In a broad sense, chemotherapy impairs mitosis by DNA damage or microtubule inhibition. Thus, mutations of genes that encode

enzymes associated with DNA replication and/or repair or dys-regulated microtubule-stabilizing proteins would result in insensitivity to chemotherapeutics.

(2) ATP-binding cassette (ABC) transporters, including breast cancer resistance protein (ABCG2), P-glycoprotein (MDR1), and multiple drug-resistant protein-1 (MRP1), have been found to be implicated in the efflux of cytotoxic remedies out of TNBC cells.

(3) Inactivation or detoxification of drugs via aldehyde dehydrogenase (ALDH), glutathione (GSH) and glutathione-S-transferase (GST) or the cytochrome P450 system, has also been demonstrated to play a part in inducing resistance in TNBC cells.

(4) Since signaling pathways are crucial in cell growth, the roles they play in chemoresistance are not negligible. It is reported that receptor-interacting protein 2 (RIP2)-mediated NF-κB activation and kinesin family member 14 (KIF14)-mediated AKT phosphorylation in TNBC promote resistance to cytotoxic drugs.

(5) Tumor microenvironment, which has a distinct impact on tumorigenicity, is found to be involved in fostering resistance. Amid tumor cells, there exists a subpopulation of cells harboring the stem-cell like properties, termed cancer stem cells (CSCs). With the ability of self-renewal, chemoresistant CSCs are capable of forming a tumor less vulnerable to chemotherapeutic agents. In addition, in the niche, hypoxia adaptation is suggested to potentially enable such unresponsiveness in TNBC. The properties of CSCs and their roles as therapeutic targets in TNBC will be discussed in more details in the next chapter.

Fate of chemoresistance

One thing to be noted is that chemotherapeutic agents promote tumor cell death through inducing cell apoptosis. Hence, altered proteins related with apoptosis such as p53, Bcl-2, and Bcl-xL have been demonstrated to be correlated with chemoresistance.

Moreover, rather than apoptosis, cells could sustain in alternative cellular fates to such as cell senescence and autophagy. In response to the pressure such as DNA damage brought about by drugs, cells turn themselves into a senescent state while remaining metabolically active.

Although no longer being able to replicate or proliferate, intriguingly, these cells could trigger their surrounding cancer cells to proliferate and transform, thereby promoting tumor progression. Aside from the senescence pathway, autophagy also contributes to cell survival following cytotoxic stresses. As is well known, autophagy can play dual roles in mediating a cell's response to chemotherapeutic drugs. On the one hand, autophagy could enhance chemotherapeutic efficacy through promoting apoptosis. On the other hand, autophagy could assist in keeping cellular viability as well. Sophisticated as autophagy is, eventually which way it goes through would rest on both the tumor and the agents [45].

6.3.3 *Targeted Therapy Using PARP Inhibitors*

Poly (ADP-ribose) polymerases (PARPs) are a family of nuclear enzymes involved in the detection and repair of DNA damage. PARPs are activated by single-strand breaks (SSBs), thus synthesizing poly (ADP-ribose) chains that serve as a signal and platform to recruit other DNA repair proteins. So far, 17 PARP members have been identified in humans [46]. PARP1 is the best-characterized member of the PARP family and is responsible for 85–90% of the total PARP activity. The reaction catalyzed by PARPs is called poly ADP-ribosylation or PARylation, which is of great importance in DNA damage repair and produces multiple cellular effects, such as DNA duplication and transcription. Failure to repair SSBs because of the defect in PARylation owing to PARP deficiency or inhibition leads to the formation of double-strand breaks (DSBs), which could be repaired by homologous recombination (HR).

As mentioned in Chapters 3 and 5, the HR-mediated repair of DSBs requires the presence of functional BRCA1 and BRCA2. Therefore, BRCA-mutated tumors are sensitive to inhibition of PARPs due to combined loss of PARP and HR repair, an effect called "synthetic lethality" [47, 48] (**Figure 6-2**). In the presence of PARP inhibitors (PARPi's), cells with BRCA defects cannot repair the DNA damage and die, whereas cells with functional BRCAs (BRCA$^{+/+}$ or BRCA$^{+/-}$) could perform effective DNA damage repair and survive.

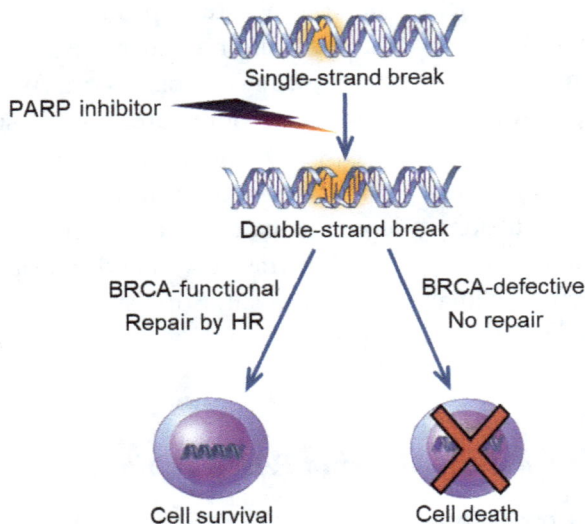

Figure 6-2. Synthetic lethality induced by PARPi's and BRCA deficiency. Exposure of cells to PARPi's leads to an accumulation of SSBs that ultimately results in the formation of DSBs. Cells with intact BRCA function (BRCA$^{+/+}$ or BRCA$^{+/-}$) could survive since these breaks can be repaired by HR, while cells with BRCA deficiency (BRCA$^{-/-}$) die since DSBs cannot be repaired. This phenomenon is known as "synthetic lethality". PARP, poly (ADP-ribose) polymerase; PARPi's, poly (ADP-ribose) polymerase inhibitors; SSB, single-strand break; DSB, double-strand break; HR, homologous recombination.

As described in the previous chapter, PARPs can be used as biomarkers for TNBC. These proteins also serve as excellent targets that can be used in the treatment of TNBC. In this regard, use of PARPi's as a therapeutic strategy in the treatment of TNBC, including basal-like phenotype, may be a promising approach. Various PARPi's have been developed to hamper DNA repair by blocking PARP-mediated PARylation. Up to now, four PARPi's, i.e., olaparib, rucaparib, niraparib, and talazoparib, have been approved by the FDA for cancer treatment. Two of them, olaparib and talazoparib, have been approved for BRCA-mutated metastatic breast cancer [49]. Another PARPi, veliparib, is being actively investigated in clinical trials for HER2-negative breast cancer or TNBC. The

Table 6-4. PARPi's and their development stages.

PARPi	Development stage	Type of cancer
Olaparib	FDA-approved	Ovarian cancer (BRCA-mutated or platinum-sensitive) Breast cancer (BRCA-mutated HER2-negative)
Talazoparib	FDA-approved	Breast cancer (advanced, BRCA-mutated, HER2-negative)
Rucaparib	FDA-approved	Ovarian cancer (advanced, germline and somatic BRCA-mutated)
Niraparib	FDA-approved	Ovarian cancer (unselected platinum-sensitive)
Veliparib (ABT-888)	Clinical trials	Breast cancer (HER2-negative or triple-negative)

Ref: *Curr Treat Options Oncol. 2018; 19(5):21.*

different PARPi's and their development stages are summarized in **Table 6-4**.

While PARPi's have been evaluated in clinical trials for TNBC as monotherapies, combination of PARPi's with DNA-damaging chemotherapy appears to be a more promising approach both to increase efficacy of PARPi's in BRCA-mutated breast cancer and to sensitize wild-type BRCA patients. Furthermore, it has been demonstrated that PARP inhibition potentiates the effects of DNA-methylating compounds, topoisomerase inhibitors, and ionizing radiation [50]. Olaparib has been combined with paclitaxel, cisplatin, or carboplatin in BRCA-mutated or metastatic TNBC [51, 52]. Veliparib has been extensively studied in combination with various chemotherapeutic drugs. In a phase I clinical trial, the combination of veliparib plus cisplatin and vinorelbine (a microtubule-destabilizing agent) gave rise to an overall response rate of 73% and 53% in TNBC with mutated and wild-type BRCA1/2, respectively [53]. In a phase III trial [54], veliparib has been combined with paclitaxel plus carboplatin for the treatment of TNBC in standard neoadjuvant chemotherapy [55]. Therefore, PARPi's have shown great promise in TNBC patients with mutated or wild-type BRCA.

6.3.4 *Radiotherapy*

As with other subtypes, radiotherapy is also performed in patients with TNBC after surgery. Interestingly, with the increasingly used BCS followed by radiation in early-stage TNBC, the emerging data show that patients with TNBC might not benefit from radiotherapy as much as those with other breast cancer subtypes. However, as previously discussed, TNBC patients with mutant BRCA1/2 fail to repair DNA DSBs via homologous recombination, thereby rendering cancer cells more vulnerable to radiotherapy [31].

6.4 Prognosis of TNBC

High heterogeneity and lack of targeted therapies have contributed to the poor prognosis of TNBC. It has been suggested that TNBC might be regarded as an independent prognostic variable, apart from the clinicopathological factors such as nodal status, tumor size, grade, and therapeutic options [56, 57]. Therefore, compared with other subtypes of breast cancer, TNBC has poor prognosis.

It should be noted that TNBC subtyping, both histologically and molecularly as described in Chapter 2, is a crucial predictor of prognosis. In addition, there are other variables that could predict the prognosis of TNBC, including the status of several biomarkers like EGFR, genes such as BRCA1, and other personal or behavioral factors like obesity [58]. Furthermore, along with more in-depth studies that explore the relationship between immune system and TNBC, it is believed that factors associated with the patient's immune status might be an important factor to predict prognosis.

6.5 Nursing Implications

As TNBC is highly aggressive and has a high risk of recurrence, patients with TNBC always experience a distressing life, scared of relapse. Thus, those patients could be emotionally disturbed due to such an uncertainty. To provide a high-quality and patient-centered care, nursing professionals are expected to have a sound grasp of knowledge about TNBC and factors that might impact the outcome in particular. Some detailed characteristics including age, ethnicity, patient perceptions, and level of education, may serve as choices of treatment and prognosis.

What is more, as required by precision medicine, individual treatment should be taken into consideration, especially mentally. Nowadays, more and more therapies for TNBC have been developed, ranging from neoadjuvant/adjuvant chemotherapy, surgery methods to targeted therapeutics. Along with the benefit brought by various agents, however, they also make it much harder for patients to make decisions. At such stages, the support from nursing practitioners is greatly needed. Additionally, it should be noted that for racial and ethnic minority patients with TNBC, their needs and coping styles might differ.

Even when the whole course of treatment is completed, the follow-up care and survivorship still require the involvement of nursing professionals to provide support. Since the rate of relapse is higher within first three years after diagnosis, it is recommended for nurses to integrate the proposal for self-care into routine clinical practice to not only ease the anxiety, but also empower those patients with TNBC.

References

1. Sharma P. (2016). Biology and management of patients with triple-negative breast cancer. *Oncologist*, **21**(9): 1050–1062.
2. Bauer KR, Brown M, Cress RD, Parise CA, Caggiano V. (2007). Descriptive analysis of estrogen receptor (ER)-negative, progesterone receptor (PR)-negative, and HER2-negative invasive breast cancer, the so-called triple-negative phenotype: a population-based study from the California cancer Registry. *Cancer*, **109**(9): 1721–1728.
3. Dent R, Trudeau M, Pritchard KI, Hanna WM, Kahn HK, Sawka CA, Lickley LA, Rawlinson E, Sun P, Narod SA. (2007). Triple-negative breast cancer: Clinical features and patterns of recurrence. *Clin Cancer Res*, **13**(15 Pt 1): 4429–4434.
4. Xu H, Eirew P, Mullaly SC, Aparicio S. (2014). The omics of triple-negative breast cancers. *Clin Chem*, **60**(1): 122–133.

5. Foulkes WD, Smith IE, Reis-Filho JS. (2010). Triple-negative breast cancer. *N Engl J Med*, **363**(20): 1938–1948.
6. Kennecke H, Yerushalmi R, Woods R, Cheang MCU, Voduc D, Speers CH, Nielsen TO, Gelmon K. (2010). Metastatic behavior of breast cancer subtypes. *J Clin Oncol*, **28**(20): 3271–3277.
7. Colzani E, Johansson ALV, Liljegren A, Foukakis T, Clements M, Adolfsson J, Hall P, Czene K. (2014). Time-dependent risk of developing distant metastasis in breast cancer patients according to treatment, age and tumour characteristics. *Br J Cancer*, **110**:1378.
8. Dent R, Hanna WM, Trudeau M, Rawlinson E, Sun P, Narod SA. (2009). Pattern of metastatic spread in triple-negative breast cancer. *Breast Cancer Res Treat*, **115**(2): 423–428.
9. Rebecca D, M. HW, Maureen T, Ellen R, Ping S, A. NS. (2009). Time to disease recurrence in basal-type breast cancers. *Cancer*, **115**(21): 4917–4923.
10. Brewster AM, Chavez-MacGregor M, Brown P. (2014). Epidemiology, biology, and treatment of triple-negative breast cancer in women of African ancestry. *Lancet Oncol*, **15**(13): e625–e634.
11. Dietze EC, Sistrunk C, Miranda-Carboni G, O'Regan R, Seewaldt VL. (2015). Triple-negative breast cancer in African-American women: Disparities versus biology. *Nat Rev Cancer*, **15**(4): 248–254.
12. Mavaddat N, Barrowdale D, Andrulis IL, Domchek SM, Eccles D, Nevanlinna H, Ramus SJ, Spurdle A, Robson M, Sherman M *et al.* (2012). Pathology of breast and ovarian cancers among BRCA1 and BRCA2 mutation carriers: Results from the Consortium of Investigators of Modifiers of BRCA1/2 (CIMBA). *Cancer Epidemiol Biomarkers Prev*, **21**(1): 134–147.
13. Nanda R, Schumm LP, Cummings S, Fackenthal JD, Sveen L, Ademuyiwa F, Cobleigh M, Esserman L, Lindor NM, Neuhausen SL *et al.* (2005). Genetic testing in an ethnically diverse cohort of high-risk women: A comparative analysis of BRCA1 and BRCA2 mutations in American families of European and African ancestry. *JAMA*, **294**(15): 1925–1933.
14. Olopade OI, Fackenthal JD, Dunston G, Tainsky MA, Collins F, Whitfield-Broome C. (2003). Breast cancer genetics in African Americans. *Cancer*, **97**(1 Suppl): 236–245.
15. Gluz O, Liedtke C, Gottschalk N, Pusztai L, Nitz U, Harbeck N. (2009). Triple-negative breast cancer–current status and future directions. *Ann Oncol*, **20**(12): 1913–1927.

16. Anders CK, Carey LA. (2009). Biology, metastatic patterns, and treatment of patients with triple-negative breast cancer. *Clin Breast Cancer*, 9 Suppl 2:S73–81.

17. Chikarmane SA, Tirumani SH, Howard SA, Jagannathan JP, DiPiro PJ. (2015). Metastatic patterns of breast cancer subtypes: What radiologists should know in the era of personalized cancer medicine. *Clin Radiol*, 70(1): 1–10.

18. Fink KR, Fink JR. (2013). Imaging of brain metastases. *Surg Neurol Int*, 4(Suppl 4): S209–S219.

19. Lin Nancy U, Claus E, Sohl J, Razzak Abdul R, Arnaout A, Winer Eric P. (2008). Sites of distant recurrence and clinical outcomes in patients with metastatic triple-negative breast cancer. *Cancer*, 113(10): 2638–2645.

20. Ovcaricek T, Frkovic SG, Matos E, Mozina B, Borstnar S. (2011). Triple negative breast cancer — prognostic factors and survival. *Radiol Oncol*, 45(1): 46–52.

21. Gangi A, Mirocha J, Leong T, Giuliano AE. (2014). Triple-negative breast cancer is not associated with increased likelihood of nodal metastases. *Ann Surg Oncol*, 21(13): 4098–4103.

22. Allred DC, Carlson RW, Berry DA, Burstein HJ, Edge SB, Goldstein LJ, Gown A, Hammond ME, Iglehart JD, Moench S *et al.* (2009). NCCN task force report. Estrogen receptor and progesterone receptor testing in breast cancer by immunohistochemistry. *J Natl Compr Canc Netw*, 7 **Suppl 6**: S1–S21; quiz S22–23.

23. Wolff AC, Hammond MEH, Hicks DG, Dowsett M, McShane LM, Allison KH, Allred DC, Bartlett JM, Bilous M, Fitzgibbons P. (2013). Recommendations for human epidermal growth factor receptor 2 testing in breast cancer: American Society of Clinical Oncology/College of American Pathologists clinical practice guideline update. *Arch Pathol Lab Med*, 138(2): 241–256.

24. Wolff AC, Hammond ME, Hicks DG, Dowsett M, McShane LM, Allison KH, Allred DC, Bartlett JM, Bilous M, Fitzgibbons P *et al.* (2013). Recommendations for human epidermal growth factor receptor 2 testing in breast cancer: American Society of Clinical Oncology/ College of American Pathologists clinical practice guideline update. *J Clin Oncol*, 31(31): 3997–4013.

25. Dogan BE, Turnbull LW. (2012). Imaging of triple-negative breast cancer. *Ann Oncol*, **23 Suppl 6**:vi23–29.

26. Golden DI, Lipson JA, Telli ML, Ford JM, Rubin DL. (2013). Dynamic contrast-enhanced MRI-based biomarkers of therapeutic response in

triple-negative breast cancer. *J Am Med Inform Assoc*, **20**(6): 1059–1066.

27. Eom HJ, Cha JH, Choi WJ, Chae EY, Shin HJ, Kim HH. (2017). Predictive clinicopathologic and dynamic contrast-enhanced MRI findings for tumor response to neoadjuvant chemotherapy in triple-negative breast cancer. *AJR Am J Roentgenol*, **208**(6): W225–W230.

28. Zhou J, Zhan W, Chang C, Zhang X, Jia Y, Dong Y, Zhou C, Sun J, Grant EG. (2014). Breast lesions: Evaluation with shear wave elastography, with special emphasis on the "stiff rim" sign. *Radiology*, **272**(1): 63–72.

29. Dzoic Dominkovic M, Ivanac G, Kelava T, Brkljacic B. (2016). Elastographic features of triple negative breast cancers. *Eur Radiol*, **26**(4): 1090–1097.

30. Locatelli MA, Curigliano G, Eniu A. (2017). Extended adjuvant chemotherapy in triple-negative breast cancer. *Breast Care (Basel)*, **12**(3): 152–158.

31. Wahba HA, El-Hadaad HA. (2015). Current approaches in treatment of triple-negative breast cancer. *Cancer Biol Med*, **12**(2): 106–116.

32. Gangi A, Chung A, Mirocha J, Liou DZ, Leong T, Giuliano AE. (2014). Breast-conserving therapy for triple-negative breast cancer. *JAMA Surg*, **149**(3): 252–258.

33. Chen QX, Wang XX, Lin PY, Zhang J, Li JJ, Song CG, Shao ZM. (2017). The different outcomes between breast-conserving surgery and mastectomy in triple-negative breast cancer: A population-based study from the SEER 18 database. *Oncotarget*, **8**(3): 4773–4780.

34. Carey LA, Dees EC, Sawyer L, Gatti L, Moore DT, Collichio F, Ollila DW, Sartor CI, Graham ML, Perou CM. (2007). The triple negative paradox: Primary tumor chemosensitivity of breast cancer subtypes. *Clin Cancer Res*, **13**(8): 2329–2334.

35. National Comprehensive Cancer Network (NCCN) Clinical Practice Guidelines in Oncology. Breast Cancer, Version 1. 2018.

36. Yadav BS, Sharma SC, Chanana P, Jhamb S. (2014). Systemic treatment strategies for triple-negative breast cancer. *World J Clin Oncol*, **5**(2): 125–133.

37. Caparica R, Lambertini M, de Azambuja E. (2019). How I treat metastatic triple-negative breast cancer. *ESMO Open*, **4**(Suppl 2): e000504.

38. S. K, J. H, S. L. (2014). Surgical treatment of primary breast cancer in the neoadjuvant setting. *BJS*, **101**(8): 912–924.

39. Minckwitz Gv, Untch M, Blohmer J-U, Costa SD, Eidtmann H, Fasching PA, Gerber B, Eiermann W, Hilfrich J, Huober J *et al.* (2012). Definition and impact of pathologic complete response on prognosis after neoadjuvant chemotherapy in various intrinsic breast cancer subtypes. *J Clin Oncol*, **30**(15): 1796–1804.

40. Gerratana L, Fanotto V, Pelizzari G, Agostinetto E, Puglisi F. (2016). Do platinum salts fit all triple negative breast cancers? *Cancer Treat Rev*, **48**:34–41.

41. Mustacchi G, De Laurentiis M. (2015). The role of taxanes in triple-negative breast cancer: Literature review. *Drug Des Devel Ther*, **9**:4303–4318.

42. Vaidya JS, Massarut S, Vaidya HJ, Alexander EC, Richards T, Caris JA, Sirohi B, Tobias JS. (2018). Rethinking neoadjuvant chemotherapy for breast cancer. *BMJ*, **360**:j5913.

43. Stover DG, Winer EP. (2015). Tailoring adjuvant chemotherapy regimens for patients with triple negative breast cancer. *Breast*, 24 Suppl 2:S132–135.

44. Guestini F, McNamara KM, Ishida T, Sasano H. (2016). Triple negative breast cancer chemosensitivity and chemoresistance: Current advances in biomarkers indentification. *Expert Opin Ther Targets*, **20**(6): 705–720.

45. O'Reilly EA, Gubbins L, Sharma S, Tully R, Guang MH, Weiner-Gorzel K, McCaffrey J, Harrison M, Furlong F, Kell M *et al.* (2015). The fate of chemoresistance in triple negative breast cancer (TNBC). *BBA Clin*, **3**:257–275.

46. Slade D. (2019). Mitotic functions of poly(ADP-ribose) polymerases. *Biochem Pharmacol*, **167**:33–43.

47. Ashworth A. (2008). A synthetic lethal therapeutic approach: Poly(ADP) ribose polymerase inhibitors for the treatment of cancers deficient in DNA double-strand break repair. *J Clin Oncol*, **26**(22): 3785–3790.

48. Telli ML, Ford JM. (2010). Novel treatment approaches for triple-negative breast cancer. *Clin Breast Cancer*, 10 Suppl 1:E16–22.

49. Zimmer AS, Gillard M, Lipkowitz S, Lee JM. (2018). Update on PARP inhibitors in breast cancer. *Curr Treat Options Oncol*, **19**(5): 21.

50. Curtin NJ. (2005). PARP inhibitors for cancer therapy. *Expert Rev Mol Med*, 7(4): 1–20.

51. Dent RA, Lindeman GJ, Clemons M, Wildiers H, Chan A, McCarthy NJ, Singer CF, Lowe ES, Watkins CL, Carmichael J. (2013). Phase I trial of the oral PARP inhibitor olaparib in combination with paclitaxel

for first- or second-line treatment of patients with metastatic triple-negative breast cancer. *Breast Cancer Res*, 15(5): R88.

52. Hartkopf AD, Stefanescu D, Wallwiener M, Hahn M, Becker S, Solomayer EF, Fehm TN, Brucker SY, Taran FA. (2014). Tumor cell dissemination to the bone marrow and blood is associated with poor outcome in patients with metastatic breast cancer. *Breast Cancer Res Treat*, 147(2): 345–351.

53. Lee A, Djamgoz MBA. (2018). Triple negative breast cancer: Emerging therapeutic modalities and novel combination therapies. *Cancer Treat Rev*, 62:110–122.

54. A Study Evaluating Safety and Efficacy of the Addition of ABT-888 Plus Carboplatin Versus the Addition of Carboplatin to Standard Chemotherapy Versus Standard Chemotherapy in Subjects With Early Stage Triple Negative Breast Cancer. doi: 10.1016/S1470-2045(18)30111-6

55. Loibl S, O'Shaughnessy J, Untch M, Sikov WM, Rugo HS, McKee MD, Huober J, Golshan M, von Minckwitz G, Maag D *et al.* (2018). Addition of the PARP inhibitor veliparib plus carboplatin or carboplatin alone to standard neoadjuvant chemotherapy in triple-negative breast cancer (BrighTNess): A randomised, phase 3 trial. *Lancet Oncol*, 19(4): 497–509.

56. Carey LA, Perou CM, Livasy CA, Dressler LG, Cowan D, Conway K, Karaca G, Troester MA, Tse CK, Edmiston S *et al.* (2006). Race, breast cancer subtypes, and survival in the carolina breast cancer study. *Jama*, 295(21): 2492–2502.

57. Hugh J, Hanson J, Cheang MC, Nielsen TO, Perou CM, Dumontet C, Reed J, Krajewska M, Treilleux I, Rupin M *et al.* (2009). Breast cancer subtypes and response to docetaxel in node-positive breast cancer: Use of an immunohistochemical definition in the BCIRG 001 trial. *J Clin Oncol*, 27(8): 1168–1176.

58. Jiralerspong S, Goodwin PJ. (2016). Obesity and Breast Cancer Prognosis: Evidence, Challenges, and Opportunities. *J Clin Oncol*, 34(35): 4203–4216.

Chapter SEVEN

Novel Therapeutic Strategies of Triple-Negative Breast Cancer

Guifei Li[1,2], Hui Yao[1,2], Shichao Yan[1,3], Shujun Fu[1,2,*],
Xiyun Deng[1,2,*], *and* Thomas J. Rosol[4,*]

Contents

*Corresponding authors: Shujun Fu, E-mail: shujunfu2020@hunnu.edu.cn; Xiyun Deng, E-mail: dengxiyunmed@hunnu.edu.cn; Thomas J. Rosol, E-mail: rosolt@ohio.edu
[1]Key Laboratory of Translational Cancer Stem Cell Research, Hunan Normal University, Changsha, Hunan, China.
[2]Departments of Pathology and Pathophysiology, Hunan Normal University School of Medicine, Changsha, Hunan, China.
[3]Department of General Surgery, The Second Xiangya Hospital, Central South University, Changsha, Hunan, China.
[4]Department of Biomedical Sciences, Ohio University, Athens, Ohio, USA.

Triple-negative breast cancer (TNBC) is a complex and aggressive subtype of breast cancer which lacks significant levels of ER, PR, and HER2, thereby making it difficult to treat clinically. As discussed in the previous chapter, TNBC treatment consists of two parts, namely, locoregional treatment including surgery and radiotherapy and systemic treatment which is based primarily on chemotherapy. Although many patients with early-stage TNBC are cured with chemotherapy, those with metastatic or recurrent disease have a median overall survival (OS) of 13–18 months with current treatment options [1]. The poor OS of TNBC has remained essentially unchanged over the past two to three decades due to its aggressive nature and lack of defined molecular targets. Therefore, targeted therapies for TNBC are urgently needed and have become an area of active research and development.

Although TNBC development involves multiple genetic and epigenetic alterations (as discussed in Chapters 3 and 4), targetable alterations are not common. This has hindered the development of successful targeted therapeutic strategies. However, important advances have been made in preclinical and clinical investigations in the management of TNBC, especially in recent years, which has largely benefited from molecular characterization of TNBC. Emerging therapeutic targets for TNBC include PARPs, tyrosine kinases (receptor and non-receptor type), androgen receptor, and more recently, immune checkpoint proteins. Key successes stem from the clinical use of PARPi's, which has been shown to significantly increase pathological complete response (pCR) in TNBC patients, especially those with BRCA1/2 mutations. Targeted therapy based on PARP inhibition has been discussed in Chapter 6. Immune checkpoint inhibition will be the topic of the next chapter. This chapter focuses other novel targeted therapies against TNBC, including strategies that target signaling molecules, epigenetic modifications, and cancer stem cells and the efforts to repurpose existing drugs for TNBC.

7.1 Targeting Signaling Molecules in TNBC

7.1.1 *Tyrosine Kinase Inhibition*

Tyrosine kinases (TKs) are protein kinases that transfer a phosphate group from ATP to the tyrosine residue on the substrate protein. These enzymes switch on/off intracellular signal transduction through reversible phosphorylation. Phosphorylation of proteins by TKs, in turn, is an important mechanism in communicating signals within a cell and regulating various cellular activities, such as cell growth, proliferation, differentiation, adhesion, migration, and cell death. TKs can be divided into two families, i.e., receptor TKs (RTKs) and non-receptor TKs (NRTKs).

RTKs are high-affinity cell surface receptors for many humoral factors such as growth factors, cytokines, and hormones. Of the 90 unique TK genes identified in humans, 58 encode RTKs [2]. RTKs include epidermal growth factor receptors (EGFRs) such as HER1 (EGFR), HER2, and HER3, fibroblast growth factor receptors (FGFRs) such as FGFR1 and FGFR2, platelet-derived growth factor receptors (PDGFRs) such as PDGFRA, vascular endothelial growth factor receptor (VEGFR), Kit, and Met. These RTKs are not only key regulators of normal cellular processes but also play crucial roles in the development and progression of many types of cancers. Important RTKs that have been shown to be highly expressed in TNBC include EGFR, FGFR, VEGFR, and Met. In particular, EGFR and Met have been shown to be promising therapeutic targets for TNBC due to their high expression in multiple molecular subtypes of TNBC [3]. These RTKs are under active investigation as anti-TNBC targets (discussed below).

While RTKs are responsible for transmembrane signaling, NRTKs function in signal transduction downstream of RTKs, relaying signals eventually to the nucleus to alter gene transcription. Up to now, 32 NRTKs have been identified in humans. NRTKs that have been implicated in TNBC include the Src family proteins and the Janus kinases (JAKs). Interestingly, these NRTKs are related with the cancer stem cell (CSC) properties and are therefore regarded as important anti-CSC targets in cancer treatment. In this section, only RTKs will be discussed; NRTKs as novel therapeutic targets will be discussed under Section 7.3 (Targeting Cancer Stem Cells).

7.1.1.1 *EGFR inhibition*

As mentioned in Chapter 1 and Chapter 5, overexpression of EGFR has been observed in more than half of the TNBC or basal-like breast cancers, which is correlated with a poor prognosis and decreased response to chemotherapy. EGFR activation promotes tumorigenesis and metastasis by increasing proliferation, epithelial-to-mesenchymal transition (EMT), migration, invasion, and angiogenesis. This observation has prompted a series of clinical trials for the treatment of TNBC using anti-EGFR agents, including the monoclonal antibodies, cetuximab and panitumumab, and the small-molecule EGFR inhibitors (EGFRi's), lapatinib and erlotinib [4].

Clinical data suggest a modest effect of anti-EGFR antibodies or EGFRi's as mono-therapies in TNBC. Combination therapy with other monoclonal antibodies or chemotherapeutics should be theoretically more efficacious. Unfortunately, results from clinical studies involving anti-EGFR-based combination therapy were unexpectedly unsatisfactory and were even less promising. Therefore, novel strategies for improving response to EGFR inhibition, for example, through combination of EGFRi's with other TKi's or PARPi's, are warranted. In additional effort, antibody-drug conjugates, which combine the antigen-specific targeting of monoclonal antibodies with the cytotoxicity of chemotherapeutics, have shown beneficial effect in treating metastatic TNBC in a phase II clinical study [5].

In TNBC, EGFR remains phosphorylated in the presence of EGFRi's, and persistent EGFR phosphorylation correlates with TKi resistance. Resistance to EGFR inhibition may be mediated by crosstalk between EGFR and other TK signaling molecules, such as Met. In considering the reason for the disappointing results from EGFR inhibition, it is important to bear in mind that it is likely that it is the mutation rather than the expression level of EGFR that is relevant to treatment efficacy. Indeed, in non-small cell lung cancer, two types of EGFR mutations can predict sensitivity to TKi therapy. Similar EGFR alterations have been detected in TNBC [6]. Whether these mutations can predict TKi sensitivity in TNBC is not known and needs further investigation.

7.1.1.2 *Met inhibition*

Met, the receptor for hepatocyte growth factor (HGF), activates multiple downstream effectors involved in cell survival, proliferation, and migration [4]. Met is highly expressed in TNBC cell lines and Met expression in TNBC tissues is associated with poor OS.

Unfortunately, a phase II trial of tivantinib, a small-molecule inhibitor of Met, in metastatic TNBC produced an overall response rate of only 5%, falling short of pre-specified efficacy goal [7]. It is hoped that combining Met inhibitors with other signaling pathway inhibitors will produce more favorable results.

7.1.2 *PI3K/Akt/mTOR Pathway Inhibition*

As discussed in Chapter 3, activation of the PI3K/Akt/mTOR pathway is a common event in TNBC. Ipatasertib, a highly selective Akt inhibitor, was evaluated in a phase II randomized trial in combination with paclitaxel as first-line treatment for metastatic TNBC. Ipatasertib improved progression-free survival (PFS) in the intent-to-treat population [8]. In a phase III trial, the addition of an allosteric Akt inhibitor, MK-2206, to standard chemotherapy in TNBC improved the pCR from 22.4% in the control group to 40.2% [9].

A phase I study was conducted in patients with the mesenchymal subtype of TNBC to evaluate the combination of mTOR inhibitors, temsirolimus or everolimus, with liposomal doxorubicin and bevacizumab (an angiogenesis inhibitor). Investigations were limited to those patients with aberrations in PIK3CA, Akt, or PTEN [10]. In a randomized phase II study, the addition of everolimus to cisplatin and paclitaxel did not improve pCR in stage II/III TNBC in the neoadjuvant setting [11].

7.1.3 *MAPK Pathway Inhibition*

Generally speaking, alterations in genes encoding the components of the MAPK pathway are not frequently observed in treatment-naïve TNBC. Considering the fact that EGFR is highly expressed in TNBC and can lead to upregulation of MAPK signaling [12], MAPK inhibition has been evaluated in TNBC patients in clinical studies. A randomized trial evaluating the MEK1/2 inhibitor, cobimetinib, with paclitaxel as first-line treatment for advanced TNBC showed a modest, though not statistically significant, increase in PFS [13].

7.1.4 *Androgen Receptor Inhibition*

7.1.4.1 *Androgen receptor and its intracellular signaling*

Androgens, primarily testosterone and dihydrotestosterone (DHT), are steroid hormones that regulate the development and maintenance of male characteristics, along with other functions in both men and women. In women, androgens are the precursors of estrogens. Androgens exert their actions through binding to the androgen receptor (AR), also known as nuclear receptor subfamily 3 group C member 4 (NR3C4).

AR is an androgen-inducible member of the nuclear receptor superfamily of transcription factors. In its unbound state, AR complexes with heat shock proteins (HSPs), such as HSP90, and remains in an inactive form. Upon androgen binding, AR undergoes a series of conformational changes, dissociates from HSP90, and translocates to the nucleus. In the nucleus, the AR complex binds to the androgen response element (ARE) and recruits co-regulatory activators, leading to the activation of transcription of the target genes important in CSC maintenance, such as Myc and Met [14] (**Figure 7-1**).

AR is expressed in about 70–90% of breast cancers and its expression varies between 10–50% in TNBC [15]. The proportion of TNBCs in which AR is expressed are called "luminal AR (LAR)" subtype (discussed in Chapter 2). Therefore, AR suppression by AR inhibitors represents another viable strategy for TNBC treatment. Agents that target the AR signaling pathway have the potential to allow a significant number of patients with advanced or metastatic TNBC to be treated with more effective, less toxic endocrine agents. Patients with AR-positive TNBC do not typically benefit from anti-estrogen therapy [16].

On the contrary, the proportion of TNBCs that lacks AR has been termed quadruple-negative breast cancer (QNBC) (discussed in Chapter 5). In this special subclass, several related pathway proteins are preferentially expressed that may serve as potential targets for treatment, such as ACSL4, SKP2, and EGFR [17].

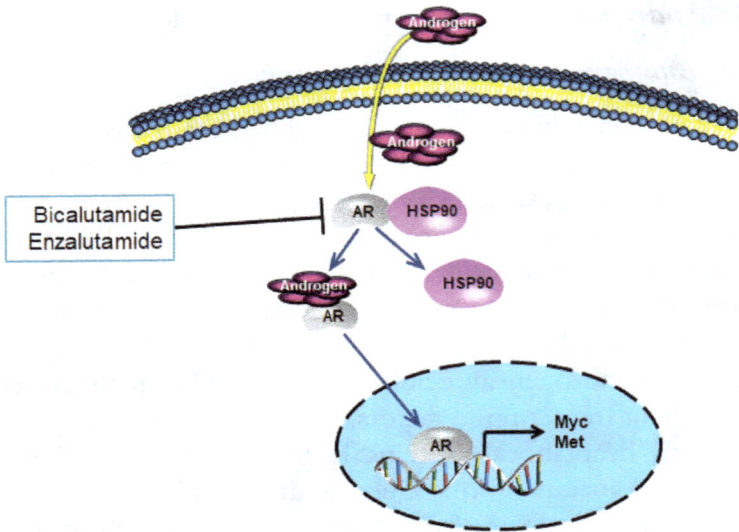

Figure 7-1. Inhibition of androgen receptor signaling pathway in TNBC. Androgens mediate their biological effects through binding to the androgen receptor (AR). In its unbound form, AR complexes with HSP90. The binding of androgens to AR induces dissociation of AR from HSP90 and subsequent receptor dimerization and translocation to the nucleus. Inside the nucleus, AR promotes gene transcription by targeting specific nucleotide palindromic sequences termed the androgen response element (ARE). Drugs that inhibit AR signaling are indicated.

7.1.4.2 *Androgen receptor signaling inhibitors*

Anti-androgens, also known as androgen antagonists, are a class of drugs that prevent androgens from mediating their biological effects in the body. They act by blocking AR or inhibiting the production of androgens. Anti-androgens include the steroidal and non-steroidal anti-androgens. Steroidal anti-androgens (e.g., cyproterone acetate) were initially developed for the treatment of prostate cancer; later, non-steroidal anti-androgens have been developed as being more effective, tolerable, and safe than the steroidal ones. The first-generation non-steroidal anti-androgens include flutamide and its derivatives, bicalutamide and nilutamide. Enzalutamide is a second-generation non-steroidal anti-androgen. The major difference between steroidal and non-steroidal anti-androgens is that the steroidal anti-androgens decrease serum levels of testosterone, whereas the non-steroidal ones do not [18].

Bicalutamide

Bicalutamide (brand name Casodex), an oral AR inhibitor, is a member of the first-generation non-steroidal anti-androgens. Bicalutamide was patented in 1982 and approved for medical use in 1995. It is the most widely used anti-androgen in the treatment of prostate cancer. A phase II study by investigators from Memorial Sloan-Kettering Cancer Center evaluated bicalutamide in patients with ER/PR-negative AR-positive breast cancer, which showed a 19% clinical benefit at 24 weeks [19]. While this drug produces side effects in men including breast enlargement, feminization, and sexual dysfunction, the side effects in women are reported to be few, except for possible harm to the baby during pregnancy and liver damage. Although the risk of adverse liver changes is small, monitoring the liver function is recommended during treatment.

Enzalutamide

Enzalutamide (brand name Xtandi), previously known as MDV3100, is a second-generation non-steroidal anti-androgen, which is more potent than the first-generation anti-androgens. Enzalutamide has at least three separate activities: 1) it functions as a potent and irreversible inhibitor of AR; 2) it impairs the translocation of AR from the cytosol into the nucleus; and 3) it blocks the interaction of AR with DNA androgen-response elements at the transcription complex [18]. Enzalutamide was first described in 2006 and approved for the treatment of prostate cancer in 2012. In patients with AR-positive metastatic TNBC, enzalutamide yielded a 42% clinical benefit at 16 weeks [20]. Based on the encouraging data available in the metastatic setting, a phase II study has been launched to evaluate the effect of enzalutamide in patients with early stage, AR-positive TNBC. Enzalutamide was found to be well tolerated at 160 mg/day. Common treatment-related adverse events include nausea, vomiting, and fatigue. In the more severe form, adverse events may include hot flashes, headache, and sexual dysfunction.

7.1.4.3 *Androgen receptor inhibition combined with other targeted drugs*

Cyclin-dependent kinase 4 (CDK4) and CDK6 play a crucial role in the G_1-S phase transition of the cell cycle. Inhibitors of CDK4/6 have been successfully used to treat hormone receptor-positive, advanced-stage breast cancer [21]. In some breast cancer subtypes, elevated expression of cyclin D1 and Rb protein was associated with increased sensitivity to CDK4/6 inhibitors such as palbociclib (trade name Ibrance). Because AR-positive TNBC has intact Rb protein, a target of palbociclib, a phase I/II clinical trial sponsored by Memorial Sloan Kettering Cancer Center in collaboration with Pfizer evaluating the combination of bicalutamide and palbociclib in women with AR-positive TNBC has been launched in November, 2015, pending release of trial results.

PI3K/Akt/mTOR signaling is necessary for cell growth and survival and has been implicated in breast cancer development. The interaction between AR and the PI3K/Akt/mTOR signaling pathway has been a particular focus of intense research in AR-positive TNBC patients as a potential therapeutic target. A multicenter phase I/II trial of enzalutamide in combination with taselisib (a PI3K inhibitor) sponsored by Vanderbilt-Ingram Cancer Center in collaboration with the NIH is evaluating this combination for the treatment of patients with advanced AR-positive TNBC.

7.1.5 *VEGF/VEGFR Signaling Pathway Inhibition*

Expression of VEGF is much higher in TNBC compared with non-TNBC [22]. VEGF exerts its endothelial cell growth-promoting action through binding to its cognate receptor, VEGFR, a receptor type TK. The angiogenesis inhibitor bevacizumab (Avastin), a monoclonal antibody which targets VEGF, has been actively investigated in patients with TNBC with encouraging results obtained. When used in combination with first-line chemotherapy, bevacizumab has consistently exhibited improved PFS and response rate in HER2-negative breast cancer. A meta-analysis of patients with HER2-negative metastatic breast cancer (n = 2,447) demonstrated that bevacizumab improved efficacy, including 1-year OS rate, both overall and in subgroups of poor-prognosis patients [23].

7.2 Targeting Epigenetic Modifications in TNBC

As discussed in Chapter 4, epigenetics is the dynamic regulation of gene expression without changes in primary nucleotide sequences. This involves a variety of regulatory mechanisms including modifications on DNA and histones and those mediated by non-coding RNAs. Such mechanisms play an important role in TNBC and agents targeting epigenetic modifications are being evaluated in TNBC.

7.2.1 *DNMT Inhibition*

DNA methyltransferases (DNMTs) are the enzymes responsible for the methylation of the promoters of genes involved in diverse cellular functions. DNMT1 is involved in aberrant methylation, and thus inactivation, of a variety of genes involved in the pathogenesis of TNBC, including BRCA1/2. Aberrant methylation of the BRCA1 promoter is found in 11–14% of all sporadic breast cancers [24, 25]. Given that BRCA1 inactivation contributes to tumorigenesis by accelerating genomic instability, BRCA1 methylation would have to be an early event that is clinically significant. It has been demonstrated that DNMT1 was elevated in TNBC at the gene expression level [26].

DNMT inhibitors (DNMTi's) are small natural or synthetic molecules able to reverse DNA hypermethylation through inhibition of DNMTs. A growing body of evidence demonstrates that plant-derived bioactive compounds with anti-cancer properties, including green tea polyphenols, soy isoflavones, curcumin, and resveratrol, also exert inhibitory effects on DNMTs. DNMTi's used as therapeutics could reactivate tumor-suppressor genes and reprogram cancer cells towards growth arrest and apoptosis [27].

Two DNMTi's that have been approved by the FDA to be used in the clinic include 5-azacytidine (Vidaza™) and its deoxy derivative, decitabine (Dacogen®, also known as 5-aza-2′-deoxycytidine). These DNMTi's have been used to treat other cancers, such as myeloid malignancies [28]. Azacytidine inhibits methylation of replicating DNA by stoichiometric binding with DNMT1, resulting in DNA hypomethylation. Decitabine functions in a similar manner to azacytidine, although decitabine incorporates into DNA strands only, while azacytidine incorporates into both DNA and RNA. This category of drugs may facilitate lower doses, thus decreasing the toxicity, of the conventional cytotoxic drugs, bring about better responses to drugs, reduce recurrence, and result in better cure rates and survival in cancer.

A phase I/II clinical trial (NCT01194908) has been launched to investigate whether decitabine has the ability to reactivate ER in TNBC patients. The rationale for this study is that ER is actually present in some TNBCs but is "silenced" because methyl groups are attached to it, rendering it in an inactive state. If ER is reactivated in the cancer cells, then these patients can be treated with anti-estrogens such as tamoxifen.

Another phase II clinical trial (NCT03295552) was initiated in November 2017 to investigate the effect of decitabine together with carboplatin for the treatment of metastatic TNBC. This trial is expected to be completed by the end of 2020.

7.2.2 HDAC Inhibition

As discussed in Chapter 4, the proper functioning of chromatin relies on reversible histone acetylation/deacetylation. In general, acetylation on histones by histone acetyltransferases (HATs) disrupts the compacted structure of chromatin to facilitate gene expression, while deacetylation by histone deacetylases (HDACs) inhibits gene expression. HDACs are involved in tumorigenesis of a variety of cancer types including TNBC. Specifically, HDACs exert their tumor-promoting effects by inhibiting the expression of tumor suppressor genes, including DNA repair genes. Two HDACi's under active investigation are vorinostat and romidepsin. HDAC inhibition by vorinostat has been shown to induce a homologous recombination deficiency-like gene expression profile in BRCA-wild-type TNBC cell lines [29]. Therefore, when combined with genotoxic agents such as cisplatin or PARPi's, HDACi's caused further reductions in cell viability in preclinical models [30]. Furthermore, in vivo data suggest that HDACi's caused TNBC cells to express ER and become sensitive to endocrine therapy [31]. Clinical trials with HDACi's in combination with aromatase inhibitors in an effort to block both HDAC and ER signaling are underway in TNBC patients [32].

Short-term single-agent treatment with vorinostat to newly diagnosed breast cancer patients was reported to decrease the expression of proliferation-associated genes, such as Ki67. Concurrent therapy of metastatic breast carcinomas with vorinostat, paclitaxel, and bevacizumab in a phase I/II Study (NCT00368875) reported a response rate of 55%.

7.3 Targeting Cancer Stem Cells in TNBC

Cancer stem cells (CSCs), which have tumor-initiating potential and are thus called tumor-initiating cells, possess self-renewal capacity and unlimited proliferative potential. CSCs are thought to be the major source of therapy resistance, metastasis, and recurrence, which are responsible for the poor outcomes across a variety of cancer types. TNBC cells have been consistently reported to display CSC signatures at functional, molecular, and transcriptional levels. In recent decades, CSC-targeting strategies have shown therapeutic potential on TNBC in multiple preclinical studies, and some of these strategies are currently being evaluated in clinical trials.

The first CSCs were discovered in human acute myelogenous leukemia (AML), in which a small population characterized by hematopoietic progenitor cell antigen CD34+ and CD38- were highly enriched and able to transfer disease [33]. Later on, many solid tumor types, including breast cancer, are found to possess a small subpopulation of replenishing stem-like cells that can give rise to the differentiated cells that comprise the bulk tumor. Unlike rapidly dividing cancer cells within the tumor mass, CSCs have a slower cycle and can survive the conventional cancer therapies that kill rapidly dividing cells. These cells are more resistant than the bulk tumor cells and, therefore, need to be more specifically targeted and eliminated in order to achieve tumor ablation [34]. This notion is beginning to revolutionize the approaches to modern cancer therapy and drug design.

7.3.1 TNBC Cancer Stem Cell Markers

In 2003, Al-Hajj first identified that the cell fraction with the phenotype characterized by positive expression of CD44 (hyaluronic acid receptor), negative or low expression of CD24 (a ligand for P-selectin), and negative expression of Lineage markers in breast cancer patient tissues had breast cancer stem cell (BCSC) characteristics [35]. As few as 1,000 CD44$^+$/CD24$^{-/low}$/Lin$^-$ cells (10–50 times lower cell number compared with unfractionated cells) can give rise to 100% tumor formation in NOD/SCID mice. Further enrichment of CD44$^+$/CD24$^{-/low}$/Lin$^-$ cell population by isolating the epithelial-specific antigen (ESA)-positive subset could further enhance tumorigenic activity in mice by about 5-fold. In 2007, Ginestier *et al.* discovered that a subpopulation of breast cancer cells with high aldehyde dehydrogenase 1 (ALDH1) activity could initiate tumors in vivo and in vitro [36]. Since then, the CD44$^+$/CD24$^{-/low}$ phenotype and high ALDH activity have become the "gold standard" signature for BCSCs. In our laboratory, enrichment by mammosphere formation in suspension culture combined with magnetic bead-assisted separation of the CD44$^+$/CD24$^{-/low}$ subset could give rise to cancer stem-like cells with 100-fold enhanced tumorigenicity compared with unfractionated cells [37].

7.3.2 Similarities Between TNBC and Breast Cancer Stem Cells

The CSC theory provides a unique insight into the aggressive nature of TNBC. Histopathological analyses of breast cancer patient tissues have revealed that compared with non-TNBC tissues, TNBC tissues exhibit enriched ALDH1 and CD44$^+$/CD24$^-$ expression signatures. Additionally, TNBC cells have been reported to form mammospheres at a higher rate than non-TNBC cells. At the transcriptional level, stemness-related transcription factors, such as Sox2 and c-Myc, are overexpressed in TNBC and are correlated with poor prognosis [38]. Microarray data from the GEO database (https://www.ncbi.nlm.nih.gov/geo) revealed that the gene signature of TNBC cells was remarkably similar to that of mammary stem cells and that the stem cell signature was significantly enriched in TNBC cell lines [38].

The similarities between TNBC and CSC phenotypes might simply be a reflection of a higher content of CSCs in TNBC. Alternatively, TNBC cells, particularly the basal-like subtype, resemble many features of BCSCs, including the CD44^{high}, CD24^{low}, and ALDH1-positive phenotype. These cells contribute to the malignant behavior such as aggressive proliferation, drug resistance, high metastatic capacity, and poor OS of TNBC patients [39]. Understanding the mechanisms that underlie the self-renewal behavior of CSCs is crucial for the discovery and development of anti-cancer agents against TNBC. Of particular importance are the signaling pathways including Wnt/β-catenin, Notch, Hedgehog, and JAK2/STAT3 pathways [40]. These pathways may play a crucial role in the recurrence and maintenance of CSCs and thus are important potential targets of TNBC.

Furthermore, evidence indicates that EMT is another similarity between TNBC and CSC phenotypes. EMT is a cellular process that promotes the conversion of adherent epithelial cells into mesenchymal-like cells. During tumor progression, EMT is thought to be activated and ultimately facilitate tumor cell migration through the basement membrane and subsequent invasion into adjacent tissues, followed by entry into the systemic circulation [41]. In particular, ectopic overexpression of EMT-promoting transcription factors, such as Snail, Twist, and Zeb1, promotes the transformation of mammary epithelial cells to BCSCs, suggesting that EMT may be a key process for the de novo generation of BCSCs [42].

EMT can be regulated by various stimuli, including oncogenic mutations and complex signaling networks that involve the tumor microenvironment; thus, EMT can serve as a key mechanism for balancing the cell state in which reprogramming to a CSC state enables the adaptation of cells to survive harsh conditions [38]. Interestingly, TNBC cells highly express the transcription factors that induce EMT, with consequent upregulation of mesenchymal proteins and downregulation of epithelial proteins. Thus, the EMT signature, which is consistently observed in both CSCs and TNBC cells, provides evidence to support the similarities between the TNBC and CSC phenotypes.

7.3.3 *Targeting Self-Renewal Capacity*

Mammary stem cells regulate their self-renewal by multiple signaling pathways that are under strict regulation by intrinsic and extrinsic mechanisms, thus maintaining homeostasis in healthy tissues. In BCSCs, the stem-like properties, including self-renewal, treatment resistance and aggressiveness, are coordinated by a network of cellular signaling pathways, including the Notch, Hedgehog, Wnt/β-catenin, and JAK/STAT3 signaling pathways [43, 44]. Alterations in one or more of these pathways have been identified in TNBC CSCs. Therefore, targeting these signaling pathways is an attractive strategy for TNBC therapy.

7.3.3.1 *JAK/STAT3 signaling inhibition*

In BCSCs, activation of JAK/STAT3 signaling has been implicated as an important mechanism for self-renewal regulation. IL6, a classic cytokine that activates the JAK/STAT3 signaling pathway, was discovered to induce the conversion of non-BCSCs into BCSCs by activating Oct4 transcription, thereby increasing the self-renewal activity of BCSCs [45]. In addition, the binding of vascular endothelial growth factor (VEGF) to its receptor 2 (VEGFR2) induces STAT3 phosphorylation and promotes the binding of STAT3 to the Sox2 and c-Myc promoter regions for their transcriptional activation in breast cancer cells [46]. Through this mechanism, VEGF-mediated STAT3 activation increases the in vivo tumorigenic potential, mammosphere-forming efficiency, and ALDH activity of BCSCs. Importantly, STAT3 overexpression was determined to be highly related to TNBC initiation, progression, metastasis, and chemotherapy resistance. In breast cancer patients, the genomic signature of TNBC conferring JAK2/STAT3 activation, which is required for growth of CD44+/CD24− stem cell-like breast cancer cells, was a predictive tool for poor prognosis [47].

There are various strategies for blocking STAT3 signaling. First, the ligand-receptor interaction can be blocked by antibodies. Second, STAT3 phosphorylation can be inhibited by targeting the activity of its upstream kinases, such as JAK, or by interfering with the docking of STAT3 to its kinases. Third, the transcriptional activity of STAT3 can be blocked by preventing its dimerization, nuclear trafficking, and binding to DNA.

Ruxolitinib, an ATP-competitive inhibitor of JAK1/JAK2, is currently in a phase II trial as a preoperative chemotherapy approach for inflammatory TNBC and in a phase I trial as a combinatorial drug for metastatic TNBC.

7.3.3.2 *Src kinase inhibition*

Src family kinases (SFKs) are members of the NRTK family that regulate signal transduction by interacting with a diverse set of cell surface receptors under multiple cellular conditions [48]. SFKs play critical roles in cell adhesion, invasion, proliferation, and survival during tumor development [49]. SFKs have been implicated in the maintenance of self-renewal capacity and chemo-/radioresistance of BCSCs. Src is more highly phosphorylated in mammospheres than in monolayer-cultured breast cancer cells, suggesting that BCSCs exhibit higher Src kinase activity. Therefore, targeting SFKs might represent a promising strategy to inhibit the CSC properties of TNBC.

Small-molecule Src inhibitors that have been developed and well-studied include dasatinib, saracatinib, and bosutinib. Unfortunately, several phase II clinical trials with these inhibitors in TNBC did not produce clinical benefit [50–52]. Dasatinib is an orally available dual tyrosine kinase inhibitor that binds to the ATP-binding site of Src kinase. Dasatinib was approved for use in patients with chronic myelogenous leukemia (CML). In breast cancer, dasatinib treatment reduced the proliferation of TNBC cell lines in vitro as well as their tumorigenic potential in vivo [53]. Moreover, dasatinib treatment sensitized TNBC cells to paclitaxel by inducing apoptosis [54]. Currently, several studies are being carried out to evaluate dasatinib as monotherapy or in combination with chemotherapy in treating TNBC [55, 56]. SKI-606 is an ATP-competitive dual inhibitor for BCR-ABL and Src kinase. A phase I study of SKI-606 is currently recruiting patients with HER2-negative and other breast cancer subtypes.

7.3.3.3 *Wnt/β-catenin signaling inhibition*

The activation of Wnt/β-catenin signaling can be initiated by the binding of Wnt ligands to their receptors, the frizzled (FZD) family proteins, and their coreceptors, low-density lipoprotein receptor-related proteins (LRPs), resulting in Wnt-FZD-LRP complex formation. Subsequently, β-catenin is released from the APC complex, including the AXIN-GSK3-CK1 complex, which phosphorylates and degrades β-catenin. This active form of β-catenin translocates into the nucleus to regulate gene transcription by binding with multiple transcription factors, including lymphoid enhancer-binding factor (LEF), T cell factor (TCF), and cAMP response element-binding protein (CREB)-binding protein (CBP). AXIN-GSK3-CK1 is recruited to the Wnt-FZD-LRP complex and then phosphorylates LRPs to stabilize the structure, thus amplifying Wnt/β-catenin signaling [57].

The importance of Wnt/β-catenin signaling in BCSCs has been validated in various experimental models. A set of genes involved in Wnt/β-catenin signaling, including Wnt1, FZD1, TCF4, and LEF1, were found to be upregulated in BCSC-enriched mammospheres or in the ALDH-high subpopulation relative to that in more differentiated bulk cancer cells [58, 59]. When treated with a small-molecule Wnt/β-catenin inhibitor, BCSCs exhibited greater growth inhibition than did bulk tumor cells. Moreover, Wnt/β-catenin inhibition sensitized BCSCs to docetaxel by reducing their self-renewal activities. Genetic silencing of Wnt1 in breast cancer cells reduced the self-renewal activity and invasive potential of BCSCs, resulting in the reduction in tumorigenesis and metastasis in orthotropic xenografts in mice.

The nuclear accumulation of β-catenin, characteristic of Wnt/β-catenin signaling activation, which is evidently increased in TNBC compared with that in non-TNBC, has been shown to promote cell migration, colony formation, stem-like features, and chemoresistance of TNBC cells and drive TNBC tumorigenesis in mouse cancer models, suggesting that Wnt/β-catenin signaling is a major driving force of TNBC tumorigenesis [60]. Furthermore, upregulation of Wnt pathway genes, including CBP, FZDs, and LRPs, was observed in TNBC tumor tissues [61, 62]. In addition, Wnt coreceptors, LRPs, have also been shown to be upregulated more frequently in TNBC, favoring cell proliferation, migration, invasion, and tumor growth. Increased LRP6 expression was found in TNBC cell lines and patient tissues compared with their non-TNBC counterparts [63, 64]. In addition to LRP6, LRP8 may play an important role in TNBC CSCs and targeting of LRP8 inhibited BCSCs in TNBC [65].

There are several strategies to block Wnt/β-catenin signaling based on the location of the targets, including nuclear transcription, extracellular ligand secretion, and signaling receptors. A phase I clinical trial of LGK-974, a small-molecule inhibitor of Wnt secretion, is currently recruiting patients with multiple solid cancer types, including TNBC.

A Wnt pathway antibody, vantictumab (OMP-18R5), which was initially isolated by its ability to bind to FZD7, was later discovered to bind to FZD2, FZD5, and FZD8. Vantictumab treatment reduced tumorigenesis in multiple types of human tumor xenografts and inhibited sphere-forming efficiency [66]. A phase I study of vantictumab in combination with paclitaxel treatment has been completed in metastatic breast cancer.

Protein tyrosine kinase 7 (PTK7) is a novel LRP6-interacting protein that physically interacts with LRP6 at its transmembrane domain and maintains LRP6 protein stability, thereby enhancing the Wnt/β-catenin pathway. A PTK7-targeted antibody-drug conjugate (PTK7-ADC) was developed, which was able to deplete

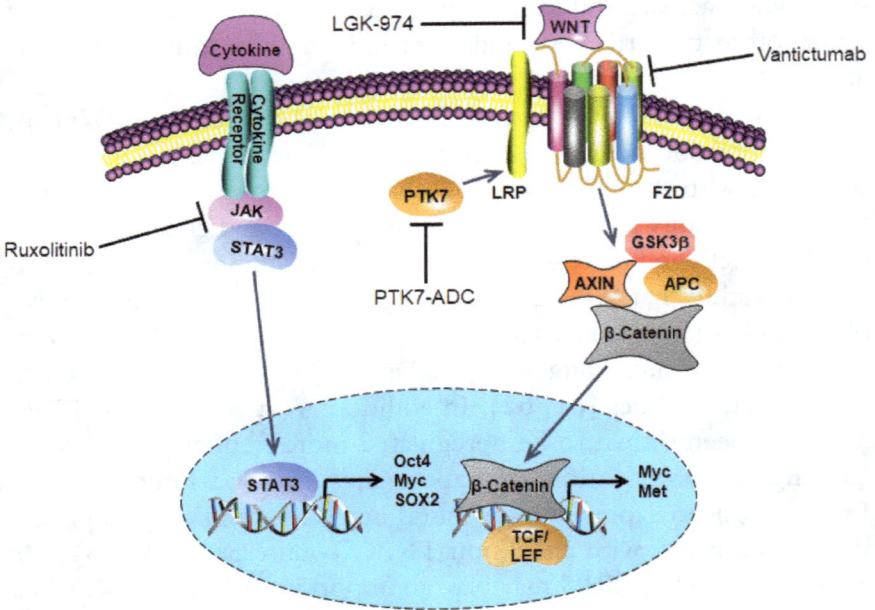

Figure 7-2. Summary of the strategies to target the self-renewal capacity of TNBC cancer stem cells. More details are described in the text. APC, adenomatous polyposis coli; AXIN, axis inhibitor; CK1, casein kinase 1; FZD, frizzled; GSK3β, glycogen synthase kinase-3β; LEF, lymphoid enhancer-binding factor; LRP, low-density lipoprotein receptor-related protein; PTK7, protein tyrosine kinase 7; PTK7-ADC, PTK7-targeted antibody-drug conjugate; TCF, T cell factor.

tumor-initiating cells and induce tumor regression in TNBC patient tissue-derived or TNBC cell line-derived xenografts [67]. Currently, a phase I study of PTK7-ADC as a combinatorial agent is recruiting patients with metastatic breast cancer and TNBC. The strategies for targeting the self-renewal capacity of TNBC are summarized in **Figure 7-2**.

7.3.4 *Targeting Metabolic Reprogramming*

CSCs possess the ability to survive the harsh microenvironment by obtaining energy from different sources depending on substrate availability. Accumulating evidence suggests that the preference of CSCs for glycolysis or oxidative phosphorylation (OXPHOS) is context-dependent. Secondary pathway, such as fatty acid oxidation (FAO), serves as an alternative strategy for fueling CSCs under additional energy-demanding conditions [68, 69]. Moreover, metabolites from CSCs can affect nearby cell populations, such as T cells and macrophages, to help CSCs escape immune surveillance.

7.3.4.1 *Targeting anaerobic glycolysis*

CSCs take advantage of anaerobic glycolysis over oxidative phosphorylation (OXPHOS) even in the presence of O_2, which is defined as the "Warburg effect". OXPHOS requires mitochondrial complex activation and increases O_2 consumption. Bypassing OXPHOS benefits CSCs by enabling rapid ATP synthesis and providing various types of metabolic intermediates for biosynthesis. Additionally, anaerobic glycolysis reduces the generation of reactive oxygen species (ROS) derived from the electron transport chain during OXPHOS. Accumulating evidence suggests that a set of glycolytic signatures, such as the elevation of glucose uptake, lactate production, and higher ATP content, are observed in BCSCs relative to those in bulk cancer cells [70, 71]. In addition, 2-deoxyglucose (2-DG), a widely used glycolytic inhibitor, can potently inhibit $CD44^+/CD24^-$ BCSCs by impairing mammosphere formation and in vivo tumorigenesis [72]. Moreover, glycolysis-mediated metabolites, particularly lactate, from BCSCs can affect various cellular compartments and thus alter the microenvironment to favor proliferation, migration, and immune escape of BCSCs [73, 74]. Furthermore, a high rate of tumor glycolysis limits the availability of glucose for tumor-infiltrating lymphocytes, which dampens the effector functions of T cells and impairs their trafficking and cytotoxicity [75, 76].

Evidence suggests that compared to non-TNBC cells, TNBC cells exhibit a glycolytic signature. Several key glycolytic proteins are preferentially overexpressed or activated in TNBC rather than in non-TNBC. Hexokinase 2 (HK2) is an enzyme in the first step of glycolysis and catalyzes the conversion of glucose to glucose-6-phosphate (G6P). It was found that TNBC patient tissues express HK2 at a higher level than non-TNBC patient tissues. HK2 expression in

TNBC was associated with EGFR expression and required for EGF-induced lactate production, suggesting the essential role of HK2 in glycolysis. The combinatorial inhibition of EGFR and glycolysis successfully reduced the in vitro and in vivo tumorigenesis of TNBC cells with minimal effect on non-TNBC cells [74].

Pyruvate kinase (PK), another key enzyme in glycolysis, catalyzes the dephosphorylation of phosphoenolpyruvate to pyruvate, which is the last irreversible step of glycolysis. The M1 isoform of PK (PKM1) is expressed in most adult tissues, whereas the M2 isoform (PKM2) is expressed exclusively during embryonic development. PKM2, particularly the active tetrameric form, is expressed in various types of tumor cells, including TNBC.

A novel mechanism of PKM2 activation was discovered in TNBC. In TNBC patient tissues and cell lines, Snail epigenetically silenced fructose-1,6-biphosphatase (FBP1) expression. Snail-mediated loss of FBP1 increased the concentration of FBP1, which activated PKM2 enzymatic activity, resulting in the elevation of glycolytic flux and ATP production to maintain TNBC cells even under hypoxia. Moreover, PKM2-mediated glycolysis was required for TNBC cells to sustain the CD44$^+$/CD24$^-$ phenotype and BCSC functions, including mammosphere formation and in vivo tumorigenesis [77].

Pyruvate dehydrogenase kinase 1 (PDK1) phosphorylates the pyruvate dehydrogenase (PDH) E1 subunit and inactivates the PDH enzyme complex that converts pyruvate to acetyl-coenzyme A (acetyl-CoA), thereby inhibiting pyruvate oxidation via the tricarboxylic acid (TCA) cycle to generate energy. PDK1 level was higher in CD44$^+$/CD24$^-$ BCSCs than in CD44$^-$/CD24$^+$ counterparts. Additionally, PDK1 protein and mRNA levels were elevated in MBA-MB-231 mammospheres and MDA-MB-231 cell subpopulations with high ALDH activity. Genetic silencing of PDK1 has been shown to significantly reduce mammosphere formation and in vivo tumorigenesis by MDA-MB-231 cells [78].

Although the number of glycolysis inhibitors is fairly limited, the widely used anti-diabetic drug, metformin, has been proposed as a potent anti-cancer drug due to its effects on multiple signaling pathways, including AMP-activated protein kinase (AMPK) and the mammalian target of rapamycin complex 1 (mTORC1). Indeed, recent experimental results have provided new insights into the mechanisms of action of metformin as a regulator of cellular metabolism, including decreased energy production and inhibited glucose uptake [79]. Moreover, metformin directly binds to HK2 to inhibit its enzymatic activity and induces the dissociation of HK2 from mitochondria [80].

In TNBC, metformin has been demonstrated to target glucose metabolism. Metformin is currently being evaluated in phase III clinical trials in patients with diverse cancer types, including TNBC. More details of metformin's potential usage in TNBC will be discussed below under the section of "Repurposing Old Drugs for TNBC" (Section 7.4).

Dichloroacetate (DCA) is an effective inhibitor of PDK1. In cancer cells, DCA switches glucose metabolism from aerobic glycolysis to OXPHOS. DCA increases OXPHOS and ROS production in mitochondria, which limits proliferation and increases apoptosis in cancer cells. A phase II study evaluating the use of DCA in metastatic breast cancer and lung cancer patients was terminated due to higher than expected risk/safety concerns. In addition, AR-12 has been developed as an orally bioavailable small-molecule inhibitor for PDK1, and a phase I study of AR-12 has been completed in patients with advanced or recurrent solid tumors, pending release of trial results.

7.3.4.2 *Targeting OXPHOS*

In contrast to the view that cancer cells switch to anaerobic glycolysis even in the presence of oxygen, Luo and coworkers recently demonstrated the importance of OXPHOS in BCSCs [81]. They showed that mitochondria in BCSCs do not lose their ability to carry out OXPHOS. Instead, under conditions of glycolysis inhibition, BCSCs are reprogrammed to use OXPHOS with higher ROS levels and become more reliant on antioxidant responses. Conversely, by reducing ROS levels using ROS scavengers, BCSCs can shift towards a more glycolytic phenotype instead of undergoing OXPHOS.

A possible relationship between OXPHOS and the intrinsic resistance of TNBC cells was revealed. The PI3K and mTOR signaling pathways are highly activated in TNBC cells; however, inhibitors of PI3K, mTOR, and mTORC1/2 have had only limited success because of resistance. Resistant TNBC cells exhibit stem-like properties, such as CD44 and ALDH1 positivity and mammosphere formation. TNBC cells were discovered to activate Notch signaling pathway following mTORC1/2 inhibition by increasing the expression of Notch1, JAG1, and active Notch1 intracellular domain (NICD). This Notch1 signaling activation was dependent on mitochondrial OXPHOS via the upregulation of certain mitochondrial transcription factors and the ATP synthase complex [82].

Emerging evidence shows that BCSCs can acquire a hybrid glycolysis/OXPHOS phenotype in which both glycolysis and OXPHOS can be used for energy production and biomolecular synthesis. The hybrid glycolysis/OXPHOS phenotype facilitates the metabolic plasticity of BCSCs and may be specifically associated with metastasis and therapy resistance [83].

Despite the importance of OXPHOS, adequate therapeutic strategies are not currently available. Recently, the therapeutic effect of a specific small-molecule inhibitor for mitochondrial electron transport chain complex I, IACS-010759, is being investigated in multiple types of cancer [84]. A phase I clinical trial is ongoing in patients with advanced cancer, including TNBC. ME-344 is another mitochondrial complex I inhibitor that is currently being evaluated in a phase I clinical trial in HER2-negative breast cancer patients. The strategies for targeting metabolic reprogramming in TNBC are summarized in **Figure 7-3**.

7.4 Repurposing Old Drugs for TNBC

7.4.1 *Statins*

Statins are inhibitors of 3-hydroxy-3-methylglutaryl coenzyme A (HMG-CoA) reductase, a key enzyme in the cholesterol synthesis pathway. Statins reduce the intracellular biosynthesis of cholesterol by reversibly inhibiting the conversion of HMG-CoA to mevalonate [85] (**Figure 7-4**). These lipid-lowering drugs are commonly used to

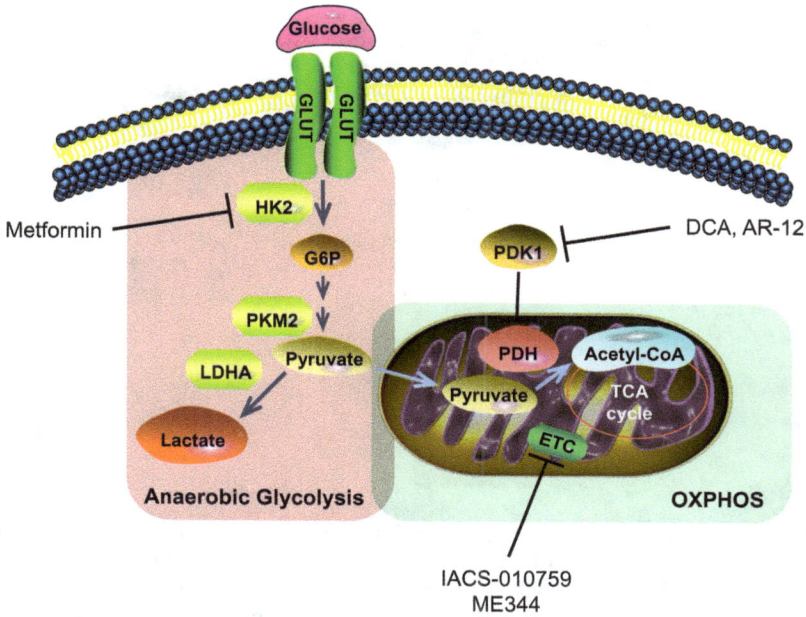

Figure 7-3. Summary of the strategies to target metabolic reprogramming in TNBC cancer stem cells. More details are described in the text. DCA, dichloroacetate; ETC, electron transport chain; GLUT, glucose transporter; LDHA, lactate dehydrogenase A; OXPHOS, oxidative phosphorylation; PKM2, pyruvate kinase M2; TCA, tricarboxylic acid.

treat hypercholesterolemia, thereby reducing the mortality from cardiovascular disease. Of great interest is the likelihood of developing cancer among the cardiovascular disease patients who use statins compared to those who do not. A meta-analysis by Group Health Research Institute at Seattle revealed that among all the medications prescribed to patients with cardiovascular disease, statins are the only class of that consistently reduce the risk of second breast cancer events [86]. This observation provides scientific support for repurposing statins for common types of cancer, including breast cancer.

Recently, statins have indeed shown pleiotropic anti-cancer effects in a variety of cancers [87]. Preclinical studies have demonstrated anti-proliferative, pro-apoptotic, anti-invasive, and radio- and

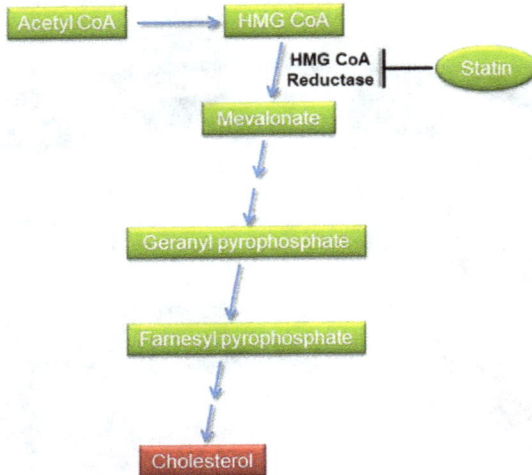

Figure 7-4. Inhibition of the cholesterol biosynthetic pathway by statins. Statins inhibit the enzyme activity of HMG-CoA reductase, which leads to ultimate reduction of the cholesterol level, through stepwise inhibition of the intermediates required for the biosynthesis of cholesterol. HMG-CoA: 3-hydroxy-3-methyl-glutaryl coenzyme A.

chemo-sensitizing properties of statins. Given that statins are FDA-approved, well tolerated, and affordable, they provide the opportunities for accelerated repurposing as cancer therapeutic agents.

7.4.1.1 *Preclinical evidence for the selective action of lipophilic statins in TNBC*

Statin sensitivity of TNBC has been associated with NFκB activation [88], lack of ERα expression [88, 89], mutation of TP53 [90], and the activity of PTEN-PI3K pathway [91]. Campbell *et al.* studied the effect of statins on the growth of breast cancer cells in vitro. Only lipophilic statins (fluvastatin, lovastatin, and simvastatin), but not a hydrophilic statin (pravastatin) significantly inhibited the proliferation of TNBC MDA-MB-231 cells with an IC_{50} less than 2 µM. However, the IC_{50} was much higher in the HER2-positive SKBr3 cells (26–49 µM) and the ER-positive MCF-7 cells (85–138 µM) [88]. Sensitivity to fluvastatin, one of the lipophilic statins, was further characterized in a panel of 19 breast cancer cell lines of various molecular subtypes and it was found that fluvastatin sensitivity was strongly associated with

Table 7-1. Sensitivity of TNBC vs. non-TNBC cell lines to lovastatin.

Subtype	Cell line	IC$_{50}$ (µM, 95% CI)	
		Normoxia	Hypoxia
TNBC	MDA-MB-231	4.7 (3.9 to 5.6)	2.5 (2.1 to 2.9)
	MDA-MB-468	12.6 (11.3 to 14.1)	5.5 (4.9 to 6.1)
	BT-549	17.2 (15.8 to 18.8)	13.4 (12.6 to 14.3)
	MX-1	1.98 (1.7 to 2.4)	7.5 (5.3 to 10.6)
Non-TNBC	MCF-7	>30 (N/A)	83.8 (48.9 to 143.8)
	T47D	>30 (N/A)	62.1 (28.2 to 136.8)
	MDA-MB-453	N/A (N/A)	>30 (N/A)

Ref: *Oncotarget 2017; 8(1):1913–1924.*

the ERα-negative status and the basal-like phenotype [92]. Xenografts of ERα-negative tumor cells have also been shown to respond to treatment with lipophilic statins [88, 90].

We extended this observation to TNBC vs. non-TNBC cell lines and confirmed that lovastatin preferentially inhibited cell proliferation of TNBC cells compared with non-TNBC cells (**Table 7-1**). A nude mouse model of orthotopic tumor growth also showed that lovastatin, at its clinically relevant concentration (2 or 10 mg/kg body weight), inhibited the in vivo tumor growth of TNBC MDA-MB-231 cells. An unappreciated molecular mechanism underlying lovastatin's selective effect on TNBC cells may involve induction of stress responses, such as nucleolar stress, through interaction with molecular target(s) distinct from its well-known lipid-lowering target. It is possible that while lipophilic statins use HMG-CoA reductase to mediate its lipid-lowering effect, they target another group of proteins, mostly likely nuclear receptors, for their nucleolar stress-inducing effect in TNBC.

7.4.1.2 *Clinical evidence supporting the use of statins as anti-TNBC agents*

Consensus regarding the clinical effects of statins on breast cancer has not been reached, which has resulted in inconsistency in the relationship between statin use and the occurrence and outcome of breast cancer. A systematic review and meta-analysis was jointly performed by Shengjing Hospital of China Medical University and Mayo Clinic and found that although statin use may not influence the risk of breast cancer, it was associated with a decrease in mortality of breast cancer patients [93]. Two further meta-analyses confirmed the overall role of statins, particularly lipophilic statins, in reducing breast cancer mortality [94, 95]. In order to ascertain whether statins benefit TNBC patients, we conducted a systematic review of the published literature regarding statin use and the outcomes of patients with TNBC or hormone receptor-negative breast cancer. We found that lipophilic statins have the potential of improving the survival of patients with TNBC or ER-negative breast cancer and, therefore, may be used as agents for the treatment of these breast cancer subtypes. Since the number of TNBC patients using statins is small (869 cases), larger-scale clinical investigations on the use of lipophilic statins and TNBC are warranted.

7.4.1.3 *Factors affecting statins' use as anti-TNBC agents*

Although statins have the potential of being repurposed for the treatment of TNBC, several factors affect their effects on TNBC. First, the lipophilicity of statins affects their potency in cancer treatment. Only lipophilic statins are able to permeate the cell membrane and affect cellular functions. This has been demonstrated in the study by Mueck *et al.* in which all lipophilic statins, i.e., lovastatin, atorvastatin, fluvastatin, and simvastatin, but not a hydrophilic statin, i.e., pravastatin, significantly inhibited the cell proliferation of breast cancer cell lines [89]. A meta-analysis by Liu *et al.* showed that lipophilic statins (but not hydrophilic ones) were associated with decreased cancer-specific and all-cause mortality in TNBC patients [95]. We confirmed this observation by showing that all lipophilic statins similarly induced nucleolar stress in TNBC CSCs [unpublished data]. Second, the duration of statin use impacts the outcome of patients with TNBC. The same group reported that while less than 4 years of statin use had a protective effect against cancer-specific and all-cause mortality in TNBC, whereas greater than 4 years of statin use did not show a beneficial effect [95]. Thirdly, the timing of statin use is another crucial determinant of its effect. Although there is no systematic study or meta-analysis on the relationship between the timing of statin use and the outcome of TNBC, it seems that the use of lipophilic statins may have a beneficial treatment effect; in contrast, they have an inconsistent or inconclusive effect on prevention of breast cancer.

7.4.2 *Metformin*

AMP-activated protein kinase (AMPK), a crucial metabolic sensor that regulates energy homeostasis at the cellular and whole-body level, is an important link between metabolism and intracellular signaling networks. The liver kinase B1 (LKB1), a major upstream kinase of AMPK, was found to be a tumor suppressor that connects AMPK signaling to carcinogenesis [96]. It has been shown that expression of AMPK is correlated with TNM staging, distant metastasis, and Ki67 status in TNBC [97]. Further studies revealed that AMPK phosphorylation was related with higher histological grade and axillary node metastasis [98]. These results demonstrate that AMPK function is compromised in TNBC and that reactivation of AMPK might have the potential to treat TNBC.

Metformin, an AMPK activator, is a first-line drug used for the treatment of type 2 diabetes mellitus (T2DM). A large body of evidence supports the positive correlation between T2DM and enhanced breast cancer risk [99]. Therefore, patients with T2DM have reduced risk of cancer when treated with metformin. Metformin also improves the prognosis of several types of cancers, including breast cancer [100]. Metformin is found to selectively target BCSCs and prolongs remission when combined with chemotherapeutic agent, doxorubicin, in a xenograft tumor mouse model [101]. Interestingly, TNBC cells are more sensitive to metformin than non-TNBC cells in terms of apoptosis induction [102]. Similar to mifepristone (discussed below), scientists at Kunming Institute of Zoology, Chinese Academy of Sciences found that metformin inhibits TNBC stem cells through targeting KLF5 for degradation [103]. The mechanisms by which metformin inhibits TNBC by targeting AMPK include: 1) inhibition of Akt/mTOR signaling; 2) inhibition of the expression of EGFR and cyclin D1/E; 3) repression of the phosphorylation of MAPK, Src, and STAT3; and 4) targeting KLF5 for degradation.

Recently, scientists at University of Chicago found that metformin has the ability to inhibit the electron transport chain (ETC) in TNBC. Further, they found BTB and CNC homology 1 (BACH1) to be a regulator of mitochondrial metabolism and a determinant of TNBC response to metformin. Combined inhibition of BACH1 by hemin and ETC by metformin exhibited greatly enhanced effects in cell line models and PDX models of TNBC [104]. These studies highlight the importance of repurposing metformin for TNBC. However, it should be kept in mind that a high concentration of metformin is required for its inhibitory effect (in the mM range). This high concentration might be achievable in the gut but not in other organs like the breast [105, 106]. This makes metformin a less likely candidate for practical clinical application at the current stage. Therefore, other AMPK activators such as AICAR and fluoxetine [107] are worth exploring as anti-TNBC agents.

7.4.3 *Mifepristone and Derivatives*

Mifepristone, one of the essential drugs that are widely used for abortion and emergency contraception, suppresses the tumor growth of TNBC cells and PDXs. The suppressive effect was found to be due to inhibition of Krüppel-like factor 5 (KLF5), a transcription factor associated with CSC properties [108]. KLF5 is frequently overexpressed in TNBC and is an unfavorable prognostic marker associated with decreased survival of breast cancer patients. Although mifepristone displays in vivo anti-tumor efficacy in animal models, its relatively modest potency has limited its further clinical application for TNBC. One major reason for this may be that mifepristone undergoes rapid metabolism resulting in significant loss of its anti-proliferative activity. To solve this issue, Lin *et al.* designed a focused compound library by altering the sensitive metabolic region of mifepristone and identified a novel mifepristone analog compound

(FZU-00004) with a satisfactory anti-cancer profile against TNBC [109]. This novel compound may serve as a promising starting point for developing mifepristone scaffold-based clinical drug candidates for TNBC.

7.5 Summary

Considering the unfavorable prognosis and aggressive nature of TNBC, many different therapeutic targets are being investigated in preclinical and clinical studies, which are summarized in **Figure 7-5**. Although the results of many new trials are encouraging, they should be interpreted with caution. Most of the current larger scale clinical evidence, if any, is based on small phase I or phase II trials. Therefore, more larger-scale trials are needed to make definite conclusions favoring clinical decisions on TNBC treatment.

We need to keep in mind that TNBC is a heterogeneous disease, which has limited the clinical benefit of many therapeutic trials in unselected TNBC patients. For example, the benefit of PARPi's, which are the only class of targeted therapies recommended to treat recurrent or metastatic TNBC, is limited to patients with the BRCA-mutated phenotype [110]. More recently, the exciting finding of the improvement of survival of TNBC patients from PDL1/PD1-targeted immunotherapy benefited a significant portion of TNBC patients [111, 112]. However, not all TNBC patients benefit from immune checkpoint-based immunotherapy. Therefore, selection of patient groups that are more likely to respond to the individual targeted therapy will help improve the clinical benefit of novel therapeutic modalities.

Furthermore, given the initial promising results of targeted therapies, drug resistance develops following the use of many targeted monotherapies due to the upregulation of compensatory signaling pathways. Combinational use of multiple targeted therapies might circumvent this issue. Therefore, future molecular targeted therapies should aim at eliminating drug resistance and selection of the group of patients that might be more likely to benefit from these novel therapeutic options.

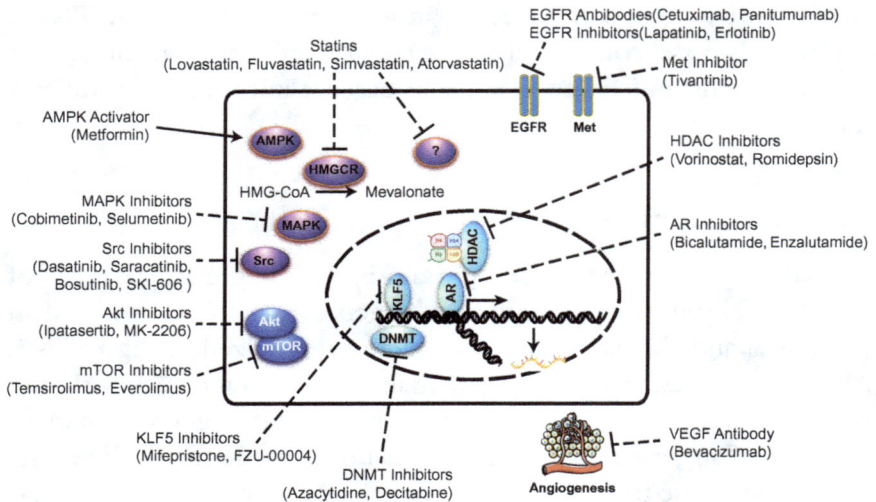

Figure 7-5. Summary of the potential agents under development for the treatment of TNBC. More details are described in the text above. Agents that target the CSC-related pathways are shown in Figure 7-2 and Figure 7-3. AMPK, AMP-activated protein kinase; AR, androgen receptor; DNMT, DNA methyltransferase; EGFR, epidermal growth factor receptor; HDAC, histone deacetylase; HMG-CoA, 3-hydroxy-3-methylglutaryl-coenzyme A; HMGCR, 3-hydroxy-3-methylglutaryl-coenzyme A reductase; KLF5, Krüppel-like factor 5; MAPK, mitogen-activated protein kinase; mTOR, mammalian target of rapamycin; VEGF, vascular endothelial growth factor.

References

1. Andre F, Zielinski CC. (2012). Optimal strategies for the treatment of metastatic triple-negative breast cancer with currently approved agents. *Ann Oncol*, **23** Suppl 6: vi46–51.

2. Robinson DR, Wu YM, Lin SF. (2000). The protein tyrosine kinase family of the human genome. *Oncogene*, **19**(49): 5548–5557.

3. Lehmann BD, Bauer JA, Chen X, Sanders ME, Chakravarthy AB, Shyr Y, Pietenpol JA. (2011). Identification of human triple-negative breast cancer subtypes and preclinical models for selection of targeted therapies. *J Clin Invest*, **121**(7): 2750–2767.

4. Lee A, Djamgoz MBA. (2018). Triple negative breast cancer: Emerging therapeutic modalities and novel combination therapies. *Cancer Treat Rev*, **62**: 110–122.

5. Burki TK. (2017). Sacituzumab govitecan activity in advanced breast cancer. *Lancet Oncol*, **18**(5): e246.

6. Cardillo TM, Sharkey RM, Rossi DL, Arrojo R, Mostafa AA, Goldenberg DM. (2017). Synthetic Lethality Exploitation by an Anti-Trop-2-SN-38 Antibody-Drug Conjugate, IMMU-132, Plus PARP Inhibitors in BRCA1/2-wild-type Triple-Negative Breast Cancer. *Clin Cancer Res*, **23**(13): 3405–3415.

7. Tolaney SM, Tan S, Guo H, Barry W, Van Allen E, Wagle N, Brock J, Larrabee K, Paweletz C, Ivanova E *et al.* (2015). Phase II study of tivantinib (ARQ 197) in patients with metastatic triple-negative breast cancer. *Invest New Drugs*, **33**(5): 1108–1114.

8. Kim SB, Dent R, Im SA, Espie M, Blau S, Tan AR, Isakoff SJ, Oliveira M, Saura C, Wongchenko MJ *et al.* (2017). Ipatasertib plus paclitaxel versus placebo plus paclitaxel as first-line therapy for metastatic triple-negative breast cancer (LOTUS): a multicentre, randomised, double-blind, placebo-controlled, phase 2 trial. *Lancet Oncol*, **18**(10): 1360–1372.

9. Biswas B. (2015). Erlotinib versus docetaxel as second- or third-line therapy in patients with advanced non-small-cell lung cancer in the era of personalized medicine. *J Clin Oncol*, **33**(5): 524.

10. Basho RK, Gilcrease M, Murthy RK, Helgason T, Karp DD, Meric-Bernstam F, Hess KR, Herbrich SM, Valero V, Albarracin C *et al.* (2017). Targeting the PI3K/AKT/mTOR Pathway for the Treatment of Mesenchymal Triple-Negative Breast Cancer: Evidence From a Phase 1 Trial of mTOR Inhibition in Combination With Liposomal Doxorubicin and Bevacizumab. *JAMA Oncol*, **3**(4): 509–515.

11. Jovanovic B, Mayer IA, Mayer EL, Abramson VG, Bardia A, Sanders ME, Kuba MG, Estrada MV, Beeler JS, Shaver TM *et al.* (2017). A Randomized Phase II Neoadjuvant Study of Cisplatin, Paclitaxel With or Without Everolimus in Patients with Stage II/III Triple-Negative Breast Cancer (TNBC): Responses and Long-term Outcome Correlated with Increased Frequency of DNA Damage Response Gene Mutations, TNBC Subtype, AR Status, and Ki67. *Clin Cancer Res*, **23**(15): 4035–4045.

12. Duncan JS, Whittle MC, Nakamura K, Abell AN, Midland AA, Zawistowski JS, Johnson NL, Granger DA, Jordan NV, Darr DB *et al.* (2012). Dynamic reprogramming of the kinome in response to targeted MEK inhibition in triple-negative breast cancer. *Cell*, **149**(2): 307–321.

13. Brufsky A, Miles D, Zvirbule Z, Eniu A, Lopez-Miranda E, Seo JH, Orditura M, Le Du F, Wongchenko M, Poulin-Costello M *et al.* (2018). Cobimetinib combined with paclitaxel as first-line treatment for patients with advanced triple-negative breast cancer (COLET study): Primary analysis of cohort I. *Cancer Res*, **78**(4).

14. Giovannelli P, Di Donato M, Galasso G, Di Zazzo E, Bilancio A, Migliaccio A. (2018). The Androgen Receptor in Breast Cancer. *Front Endocrinol (Lausanne)*, **9**: 492.

15. Gerratana L, Basile D, Buono G, De Placido S, Giuliano M, Minichillo S, Coinu A, Martorana F, De Santo I, Del Mastro L *et al.* (2018). Androgen receptor in triple negative breast cancer: A potential target for the targetless subtype. *Cancer Treat Rev*, **68**: 102–110.

16. Gucalp A, Traina TA. (2016). Targeting the androgen receptor in triple-negative breast cancer. *Curr Probl Cancer*, **40**(2–4): 141–150.

17. Hon JD, Singh B, Sahin A, Du G, Wang J, Wang VY, Deng FM, Zhang DY, Monaco ME, Lee P. (2016). Breast cancer molecular subtypes: from TNBC to QNBC. *Am J Cancer Res*, **6**(9): 1864–1872.

18. Vasaitis TS, Njar VC. (2010). Novel, potent anti-androgens of therapeutic potential: recent advances and promising developments. *Future Med Chem*, **2**(4): 667–680.

19. Gucalp A, Tolaney S, Isakoff SJ, Ingle JN, Liu MC, Carey LA, Blackwell K, Rugo H, Nabell L, Forero A *et al.* (2013). Phase II trial of bicalutamide in patients with androgen receptor-positive, estrogen receptor-negative metastatic Breast Cancer. *Clin Cancer Res*, **19**(19): 5505–5512.

20. Traina TA, Miller K, Yardley DA, Eakle J, Schwartzberg LS, O'Shaughnessy J, Gradishar W *et al.* (2018). Enzalutamide for the treatment of androgen receptor-expressing triple-negative breast cancer. *J Clin Oncol*, **36**(9):884–890.

21. Liu M, Liu H, Chen J. (2018). Mechanisms of the CDK4/6 inhibitor palbociclib (PD 0332991) and its future application in cancer treatment (Review). *Oncol Rep*, **39**(3): 901–911.

22. Linderholm BK, Hellborg H, Johansson U, Elmberger G, Skoog L, Lehtio J, Lewensohn R. (2009). Significantly higher levels of vascular endothelial growth factor (VEGF) and shorter survival times for patients with primary operable triple-negative breast cancer. *Ann Oncol*, **20**(10): 1639–1646.

23. Miles DW, Dieras V, Cortes J, Duenne AA, Yi J, O'Shaughnessy J. (2013). First-line bevacizumab in combination with chemotherapy for

HER2-negative metastatic breast cancer: pooled and subgroup analyses of data from 2447 patients. *Ann Oncol*, **24**(11): 2773–2780.

24. Esteller M, Silva JM, Dominguez G, Bonilla F, Matias-Guiu X, Lerma E, Bussaglia E, Prat J, Harkes IC, Repasky EA *et al.* (2000). Promoter hypermethylation and BRCA1 inactivation in sporadic breast and ovarian tumors. *J Natl Cancer Inst*, **92**(7): 564–569.

25. Catteau A, Harris WH, Xu CF, Solomon E. (1999). Methylation of the BRCA1 promoter region in sporadic breast and ovarian cancer: correlation with disease characteristics. *Oncogene*, **18**(11): 1957–1965.

26. Shin E, Lee Y, Koo JS. (2016). Differential expression of the epigenetic methylation-related protein DNMT1 by breast cancer molecular subtype and stromal histology. *J Transl Med*, **14**: 87.

27. Supic G ZK, and Magic Z. (2018). Chapter 5: Epigenetic Therapy of Cancer — From Mechanisms to Clinical Utility, in book: Epigenetics in cancer, Editors: Fu J, Imani S. *Narosa Publishing House/Science Press* 2018.

28. Jhan JR, Andrechek ER. (2017). Triple-negative breast cancer and the potential for targeted therapy. *Pharmacogenomics*, **18**(17): 1595–1609.

29. Peng G, Chun-Jen Lin C, Mo W, Dai H, Park YY, Kim SM, Peng Y, Mo Q, Siwko S, Hu R *et al.* (2014). Genome-wide transcriptome profiling of homologous recombination DNA repair. *Nat Commun*, **5**: 3361.

30. Min A, Im SA, Kim DK, Song SH, Kim HJ, Lee KH, Kim TY, Han SW, Oh DY, Kim TY *et al.* (2015). Histone deacetylase inhibitor, suberoylanilide hydroxamic acid (SAHA), enhances anti-tumor effects of the poly (ADP-ribose) polymerase (PARP) inhibitor olaparib in triple-negative breast cancer cells. *Breast Cancer Res*, **17**: 33.

31. Sabnis GJ, Goloubeva O, Chumsri S, Nguyen N, Sukumar S, Brodie AM. (2011). Functional activation of the estrogen receptor-alpha and aromatase by the HDAC inhibitor entinostat sensitizes ER-negative tumors to letrozole. *Cancer Res*, **71**(5): 1893–1903.

32. Entinostat and Anastrozole in Treating Postmenopausal Women With TNBC That Can Be Removed by Surgery. [https://clinicaltrials.gov/ct2/show/NCT00053898]

33. Bonnet D, Dick JE. (1997). Human acute myeloid leukemia is organized as a hierarchy that originates from a primitive hematopoietic cell. *Nat Med*, **3**(7): 730–737.

34. Batlle E, Clevers H. (2017). Cancer stem cells revisited. *Nat Med*, 23(10): 1124–1134.

35. Al-Hajj M, Wicha MS, Benito-Hernandez A, Morrison SJ, Clarke MF. (2003). Prospective identification of tumorigenic breast cancer cells. *Proc Natl Acad Sci USA*, 100(7): 3983–3988.

36. Ginestier C, Hur MH, Charafe-Jauffret E, Monville F, Dutcher J, Brown M, Jacquemier J, Viens P, Kleer CG, Liu S *et al.* (2007). ALDH1 is a marker of normal and malignant human mammary stem cells and a predictor of poor clinical outcome. *Cell Stem Cell*, 1(5): 555–567.

37. Song L, Tao X, Lin L, Chen C, Yao H, He G, Zou G, Cao Z, Yan S, Lu L *et al.* (2018). Cerasomal Lovastatin Nanohybrids for Efficient Inhibition of Triple-Negative Breast Cancer Stem Cells To Improve Therapeutic Efficacy. *ACS Appl Mater Interfaces*, 10(8): 7022–7030.

38. Park SY, Choi JH, Nam JS. (2019). Targeting Cancer Stem Cells in Triple-Negative Breast Cancer. *Cancers (Basel)*, 11(7):965.

39. O'Conor CJ, Chen T, Gonzalez I, Cao D, Peng Y. (2018). Cancer stem cells in triple-negative breast cancer: a potential target and prognostic marker. *Biomark Med*, 12(7): 813–820.

40. Liu S, Wicha MS. (2010). Targeting breast cancer stem cells. *J Clin Oncol*, 28(25): 4006–4012.

41. Kalluri R, Weinberg RA. (2009). The basics of epithelial-mesenchymal transition. *J Clin Invest*, 119(6): 1420–1428.

42. Mani SA, Guo W, Liao MJ, Eaton EN, Ayyanan A, Zhou AY, Brooks M, Reinhard F, Zhang CC, Shipitsin M *et al.* (2008). The epithelial-mesenchymal transition generates cells with properties of stem cells. *Cell*, 133(4): 704–715.

43. Takebe N, Miele L, Harris PJ, Jeong W, Bando H, Kahn M, Yang SX, Ivy SP. (2015). Targeting Notch, Hedgehog, and Wnt pathways in cancer stem cells: clinical update. *Nat Rev Clin Oncol*, 12(8): 445–464.

44. Pires BR, IS DEA, Souza LD, Rodrigues JA, Mencalha AL. (2016). Targeting Cellular Signaling Pathways in Breast Cancer Stem Cells and its Implication for Cancer Treatment. *Anticancer Res*, 36(11): 5681–5691.

45. Kim SY, Kang JW, Song X, Kim BK, Yoo YD, Kwon YT, Lee YJ. (2013). Role of the IL-6-JAK1-STAT3-Oct-4 pathway in the conversion of non-stem cancer cells into cancer stem-like cells. *Cell Signal*, 25(4): 961–969.

46. Zhao D, Pan C, Sun J, Gilbert C, Drews-Elger K, Azzam DJ, Picon-Ruiz M, Kim M, Ullmer W, El-Ashry D *et al.* (2015). VEGF drives cancer-initiating stem cells through VEGFR-2/Stat3 signaling to upregulate Myc and Sox2. *Oncogene*, **34**(24): 3107–3119.

47. Marotta LL, Almendro V, Marusyk A, Shipitsin M, Schemme J, Walker SR, Bloushtain-Qimron N, Kim JJ, Choudhury SA, Maruyama R *et al.* (2011). The JAK2/STAT3 signaling pathway is required for growth of CD44(+)CD24(-) stem cell-like breast cancer cells in human tumors. *J Clin Invest*, **121**(7): 2723–2735.

48. Parsons SJ, Parsons JT. (2004). Src family kinases, key regulators of signal transduction. *Oncogene*, **23**(48): 7906–7909.

49. Kim LC, Song L, Haura EB. (2009). Src kinases as therapeutic targets for cancer. *Nat Rev Clin Oncol*, **6**(10): 587–595.

50. Campone M, Bondarenko I, Brincat S, Hotko Y, Munster PN, Chmielowska E, Fumoleau P, Ward R, Bardy-Bouxin N, Leip E *et al.* (2012). Phase II study of single-agent bosutinib, a Src/Abl tyrosine kinase inhibitor, in patients with locally advanced or metastatic breast cancer pretreated with chemotherapy. *Ann Oncol*, **23**(3): 610–617.

51. Gucalp A, Sparano JA, Caravelli J, Santamauro J, Patil S, Abbruzzi A, Pellegrino C, Bromberg J, Dang C, Theodoulou M *et al.* (2011). Phase II trial of saracatinib (AZD0530), an oral SRC-inhibitor for the treatment of patients with hormone receptor-negative metastatic breast cancer. *Clin Breast Cancer*, **11**(5): 306–311.

52. Finn RS, Bengala C, Ibrahim N, Roche H, Sparano J, Strauss LC, Fairchild J, Sy O, Goldstein LJ. (2011). Dasatinib as a single agent in triple-negative breast cancer: results of an open-label phase 2 study. *Clin Cancer Res*, **17**(21): 6905–6913.

53. Qian XL, Zhang J, Li PZ, Lang RG, Li WD, Sun H, Liu FF, Guo XJ, Gu F, Fu L. (2017). Dasatinib inhibits c-src phosphorylation and prevents the proliferation of Triple-Negative Breast Cancer (TNBC) cells which overexpress Syndecan-Binding Protein (SDCBP). *PloS One*, **12**(1): e0171169.

54. Tian J, Raffa FA, Dai M, Moamer A, Khadang B, Hachim IY, Bakdounes K, Ali S, Jean-Claude B, Lebrun JJ. (2018). Dasatinib sensitises triple negative breast cancer cells to chemotherapy by targeting breast cancer stem cells. *Br J Cancer*, **119**(12): 1495–1507.

55. Yadav BS, Sharma SC, Chanana P, Jhamb S. (2014). Systemic treatment strategies for triple-negative breast cancer. *World J Clin Oncol*, **5**(2): 125–133.

56. Brower V. (2009). Search for new treatments intensifies for triple-negative breast cancer. *J Natl Cancer Inst*, **101**(22): 1536–1537.

57. MacDonald BT, He X. (2012). Frizzled and LRP5/6 receptors for Wnt/beta-catenin signaling. *Cold Spring Harb Perspect Biol*, **4**(12): a007880.

58. Jang GB, Hong IS, Kim RJ, Lee SY, Park SJ, Lee ES, Park JH, Yun CH, Chung JU, Lee KJ *et al.* (2015). Wnt/beta-Catenin Small-Molecule Inhibitor CWP232228 Preferentially Inhibits the Growth of Breast Cancer Stem-like Cells. *Cancer Res*, **75**(8): 1691–1702.

59. Jang GB, Kim JY, Cho SD, Park KS, Jung JY, Lee HY, Hong IS, Nam JS. (2015). Blockade of Wnt/beta-catenin signaling suppresses breast cancer metastasis by inhibiting CSC-like phenotype. *Sci Rep*, **5**: 12465.

60. Xu J, Prosperi JR, Choudhury N, Olopade OI, Goss KH. (2015). beta-Catenin is required for the tumorigenic behavior of triple-negative breast cancer cells. *PloS One*, **10**(2): e0117097.

61. Pohl SG, Brook N, Agostino M, Arfuso F, Kumar AP, Dharmarajan A. (2017). Wnt signaling in triple-negative breast cancer. *Oncogenesis*, **6**(4): e310.

62. Yang L, Wu X, Wang Y, Zhang K, Wu J, Yuan YC, Deng X, Chen L, Kim CC, Lau S *et al.* (2011). FZD7 has a critical role in cell proliferation in triple negative breast cancer. *Oncogene*, **30**(43): 4437–4446.

63. Liu CC, Prior J, Piwnica-Worms D, Bu G. (2010). LRP6 overexpression defines a class of breast cancer subtype and is a target for therapy. *Proc Natl Acad Sci USA*, **107**(11): 5136–5141.

64. Ma J, Lu W, Chen D, Xu B, Li Y. (2017). Role of Wnt Co-Receptor LRP6 in Triple Negative Breast Cancer Cell Migration and Invasion. *J Cell Biochem*, **118**(9): 2968–2976.

65. Lin CC, Lo MC, Moody R, Jiang H, Harouaka R, Stevers N, Tinsley S, Gasparyan M, Wicha M, Sun D. (2018). Targeting LRP8 inhibits breast cancer stem cells in triple-negative breast cancer. *Cancer Lett*, **438**: 165–173.

66. Gurney A, Axelrod F, Bond CJ, Cain J, Chartier C, Donigan L, Fischer M, Chaudhari A, Ji M, Kapoun AM *et al.* (2012). Wnt pathway inhibition via the targeting of Frizzled receptors results in decreased growth and tumorigenicity of human tumors. *Proc Natl Acad Sci USA*, **109**(29): 11717–11722.

67. Damelin M, Bankovich A, Bernstein J, Lucas J, Chen L, Williams S, Park A, Aguilar J, Ernstoff E, Charati M *et al.* (2017). A PTK7-

targeted antibody-drug conjugate reduces tumor-initiating cells and induces sustained tumor regressions. *Sci Transl Med*, **9**(372).

68. Peiris-Pages M, Martinez-Outschoorn UE, Pestell RG, Sotgia F, Lisanti MP. (2016). Cancer stem cell metabolism. *Breast Cancer Res*, **18**(1): 55.

69. Sancho P, Barneda D, Heeschen C. (2016). Hallmarks of cancer stem cell metabolism. *Br J Cancer*, **114**(12): 1305–1312.

70. Hammoudi N, Ahmed KB, Garcia-Prieto C, Huang P. (2011). Metabolic alterations in cancer cells and therapeutic implications. *Chin J Cancer*, **30**(8): 508–525.

71. Pecqueur C, Oliver L, Oizel K, Lalier L, Vallette FM. (2013). Targeting metabolism to induce cell death in cancer cells and cancer stem cells. *Int J Cell Biol*, **2013**: 805975.

72. Ciavardelli D, Rossi C, Barcaroli D, Volpe S, Consalvo A, Zucchelli M, De Cola A, Scavo E, Carollo R, D'Agostino D *et al.* (2014). Breast cancer stem cells rely on fermentative glycolysis and are sensitive to 2-deoxyglucose treatment. *Cell Death Dis*, **5**(7): e1336.

73. Colegio OR, Chu NQ, Szabo AL, Chu T, Rhebergen AM, Jairam V, Cyrus N, Brokowski CE, Eisenbarth SC, Phillips GM *et al.* (2014). Functional polarization of tumour-associated macrophages by tumour-derived lactic acid. *Nature*, **513**(7519): 559–563.

74. Lim SO, Li CW, Xia W, Lee HH, Chang SS, Shen J, Hsu JL, Raftery D, Djukovic D, Gu H *et al.* (2016). EGFR Signaling Enhances Aerobic Glycolysis in Triple-Negative Breast Cancer Cells to Promote Tumor Growth and Immune Escape. *Cancer Res*, **76**(5): 1284–1296.

75. Cascone T, McKenzie JA, Mbofung RM, Punt S, Wang Z, Xu C, Williams LJ, Bristow CA, Carugo A, Peoples MD *et al.* (2018). Increased Tumor Glycolysis Characterizes Immune Resistance to Adoptive T Cell Therapy. *Cell Metab*, **27**(5): 977–987.e4.

76. Sugiura A, Rathmell JC. (2018). Metabolic Barriers to T Cell Function in Tumors. *J Immunol*, **200**(2): 400–407.

77. Dong C, Yuan T, Wu Y, Wang Y, Fan TW, Miriyala S, Lin Y, Yao J, Shi J, Kang T *et al.* (2013). Loss of FBP1 by Snail-mediated repression provides metabolic advantages in basal-like breast cancer. *Cancer Cell*, **23**(3): 316–331.

78. Peng F, Wang JH, Fan WJ, Meng YT, Li MM, Li TT, Cui B, Wang HF, Zhao Y, An F *et al.* (2018). Glycolysis gatekeeper PDK1 reprograms breast cancer stem cells under hypoxia. *Oncogene*, **37**(8): 1062–1074.

79. Salani B, Del Rio A, Marini C, Sambuceti G, Cordera R, Maggi D. (2014). Metformin, cancer and glucose metabolism. *Endocr Relat Cancer*, **21**(6): R461–471.

80. Salani B, Marini C, Rio AD, Ravera S, Massollo M, Orengo AM, Amaro A, Passalacqua M, Maffioli S, Pfeffer U *et al.* (2013). Metformin impairs glucose consumption and survival in Calu-1 cells by direct inhibition of hexokinase-II. *Sci Rep*, **3**: 2070.

81. Luo M, Shang L, Brooks MD, Jiagge E, Zhu Y, Buschhaus JM, Conley S, Fath MA, Davis A, Gheordunescu E *et al.* (2018). Targeting Breast Cancer Stem Cell State Equilibrium through Modulation of Redox Signaling. *Cell Metab*, **28**(1): 69–86.e6.

82. Bhola NE, Jansen VM, Koch JP, Li H, Formisano L, Williams JA, Grandis JR, Arteaga CL. (2016). Treatment of Triple-Negative Breast Cancer with TORC1/2 Inhibitors Sustains a Drug-Resistant and Notch-Dependent Cancer Stem Cell Population. *Cancer Res*, **76**(2): 440–452.

83. Jia D, Park JH, Jung KH, Levine H, Kaipparettu BA. (2018). Elucidating the Metabolic Plasticity of Cancer: Mitochondrial Reprogramming and Hybrid Metabolic States. *Cells*, **7**(3):21.

84. Molina JR, Sun Y, Protopopova M, Gera S, Bandi M, Bristow C, McAfoos T, Morlacchi P, Ackroyd J, Agip AA *et al.* (2018). An inhibitor of oxidative phosphorylation exploits cancer vulnerability. *Nat Med*, **24**(7): 1036–1046.

85. Yao H, He G, Yan S, Chen C, Song L, Rosol TJ, Deng X. (2017). Triple-negative breast cancer: is there a treatment on the horizon? *Oncotarget*, **8**(1): 1913–1924.

86. Boudreau DM, Yu O, Chubak J, Wirtz HS, Bowles EJ, Fujii M, Buist DS. (2014). Comparative safety of cardiovascular medication use and breast cancer outcomes among women with early stage breast cancer. *Breast Cancer Res Treat*, **144**(2): 405–416.

87. Gazzerro P, Proto MC, Gangemi G, Malfitano AM, Ciaglia E, Pisanti S, Santoro A, Laezza C, Bifulco M. (2012). Pharmacological actions of statins: a critical appraisal in the management of cancer. *Pharmacol Rev*, **64**(1): 102–146.

88. Campbell MJ, Esserman LJ, Zhou Y, Shoemaker M, Lobo M, Borman E, Baehner F, Kumar AS, Adduci K, Marx C *et al.* (2006). Breast cancer growth prevention by statins. *Cancer Res*, **66**(17): 8707–8714.

89. Mueck AO, Seeger H, Wallwiener D. (2003). Effect of statins combined with estradiol on the proliferation of human receptor-positive and receptor-negative breast cancer cells. *Menopause*, **10**(4): 332–336.

90. Freed-Pastor WA, Mizuno H, Zhao X, Langerod A, Moon SH, Rodriguez-Barrueco R, Barsotti A, Chicas A, Li W, Polotskaia A *et al.* (2012). Mutant p53 disrupts mammary tissue architecture via the mevalonate pathway. *Cell*, **148**(1–2): 244–258.

91. Park YH, Jung HH, Ahn JS, Im YH. (2013). Statin induces inhibition of triple negative breast cancer (TNBC) cells via PI3K pathway. *Biochem Biophys Res Commun*, **439**(2): 275–279.

92. Goard CA, Chan-Seng-Yue M, Mullen PJ, Quiroga AD, Wasylishen AR, Clendening JW, Sendorek DH, Haider S, Lehner R, Boutros PC *et al.* (2014). Identifying molecular features that distinguish fluvastatin-sensitive breast tumor cells. *Breast Cancer Res Treat*, **143**(2): 301–312.

93. Wu QJ, Tu C, Li YY, Zhu J, Qian KQ, Li WJ, Wu L. (2015). Statin use and breast cancer survival and risk: a systematic review and meta-analysis. *Oncotarget*, **6**(40): 42988–43004.

94. Manthravadi S, Shrestha A, Madhusudhana S. (2016). Impact of statin use on cancer recurrence and mortality in breast cancer: A systematic review and meta-analysis. *Int J Cancer*, **139**(6): 1281–1288.

95. Liu B, Yi Z, Guan X, Zeng YX, Ma F. (2017). The relationship between statins and breast cancer prognosis varies by statin type and exposure time: a meta-analysis. *Breast Cancer Res Treat*, **164**(1): 1–11.

96. Shaw RJ, Kosmatka M, Bardeesy N, Hurley RL, Witters LA, DePinho RA, Cantley LC. (2004). The tumor suppressor LKB1 kinase directly activates AMP-activated kinase and regulates apoptosis in response to energy stress. *Proc Natl Acad Sci USA*, **101**(10): 3329–3335.

97. Huang X, Li X, Xie X, Ye F, Chen B, Song C, Tang H, Xie X. (2016). High expressions of LDHA and AMPK as prognostic biomarkers for breast cancer. *Breast (Edinburgh, Scotland)*, **30**: 39–46.

98. Hadad SM, Baker L, Quinlan PR, Robertson KE, Bray SE, Thomson G, Kellock D, Jordan LB, Purdie CA, Hardie DG *et al.* (2009). Histological evaluation of AMPK signalling in primary breast cancer. *BMC Cancer*, **9**(1): 307.

99. Tsilidis KK, Kasimis JC, Lopez DS, Ntzani EE, Ioannidis JP. (2015). Type 2 diabetes and cancer: umbrella review of meta-analyses of observational studies. *BMJ*, **350**: g7607.

100. Morales DR, Morris AD. (2015). Metformin in cancer treatment and prevention. *Annu Rev Med*, **66**:17–29.

101. Hirsch HA, Iliopoulos D, Tsichlis PN, Struhl K. (2009). Metformin selectively targets cancer stem cells, and acts together with chemotherapy to block tumor growth and prolong remission. *Cancer Res*, **69**(19): 7507–7511.

102. Liu B, Fan Z, Edgerton SM, Deng XS, Alimova IN, Lind SE, Thor AD. (2009). Metformin induces unique biological and molecular responses in triple negative breast cancer cells. *Cell Cycle*, **8**(13): 2031–2040.

103. Shi P, Liu W, Tala, Wang H, Li F, Zhang H, Wu Y, Kong Y, Zhou Z, Wang C *et al.* (2017). Metformin suppresses triple-negative breast cancer stem cells by targeting KLF5 for degradation. *Cell Discov*, **3**(1):1–13.

104. Lee J, Yesilkanal AE, Wynne JP, Frankenberger C, Liu J, Yan J, Elbaz M, Rabe DC, Rustandy FD, Tiwari P *et al.* (2019). Effective breast cancer combination therapy targeting BACH1 and mitochondrial metabolism. *Nature*, **568**(7751): 254–258.

105. Maniar K, Singh V, Chakrabarti A, Bhattacharyya R, Banerjee D. (2018). High dose targeted delivery on cancer sites and the importance of short-chain fatty acids for metformin's action: Two crucial aspects of the wonder drug. *Regul Toxicol Pharmacol*, **97**: 15–16.

106. He L, Wondisford FE. (2015). Metformin action: concentrations matter. *Cell Metab*, **21**(2): 159–162.

107. Cao W, Li J, Hao Q, Vadgama JV, Wu Y. (2019). AMP-activated protein kinase: a potential therapeutic target for triple-negative breast cancer. *Breast Cancer Res*, **21**(1): 1–10.

108. Liu R, Shi P, Nie Z, Liang H, Zhou Z, Chen W, Chen H, Dong C, Yang R, Liu S *et al.* (2016). Mifepristone Suppresses Basal Triple-Negative Breast Cancer Stem Cells by Down-regulating KLF5 Expression. *Theranostics*, **6**(4): 533–544.

109. Lin Y, Liu R, Zhao P, Ye J, Zheng Z, Huang J, Zhang Y, Gao Y, Chen H, Liu S *et al.* (2018). Discovery of novel mifepristone derivatives via suppressing KLF5 expression for the treatment of triple-negative breast cancer. *Eur J Med Chem*, **146**: 354–367.

110. Telli ML, Timms KM, Reid J, Hennessy B, Mills GB, Jensen KC, Szallasi Z, Barry WT, Winer EP, Tung NM *et al.* (2016). Homologous Recombination Deficiency (HRD) Score Predicts Response to Platinum-Containing Neoadjuvant Chemotherapy in Patients with Triple-Negative Breast Cancer. *Clin Cancer Res*, **22**(15): 3764–3773.

111. Schmid P, Adams S, Rugo HS, Schneeweiss A, Barrios CH, Iwata H, Dieras V, Hegg R, Im SA, Shaw Wright G *et al.* (2018). Atezolizumab and Nab-Paclitaxel in Advanced Triple-Negative Breast Cancer. *N Engl J Med*, **379**(22): 2108–2121.

112. Nanda R, Chow LQ, Dees EC, Berger R, Gupta S, Geva R, Pusztai L, Pathiraja K, Aktan G, Cheng JD *et al.* (2016). Pembrolizumab in Patients With Advanced Triple-Negative Breast Cancer: Phase Ib KEYNOTE-012 Study. *J Clin Oncol*, **34**(21): 2460–2467.

Chapter EIGHT

Immunotherapy of Triple-Negative Breast Cancer

Huimei Yi[1,2], Yuan Tan[3], Lu Lu[1,2], Faqing Tang[3,*], and Xiyun Deng[1,2,*]

Contents

*Corresponding authors: Faqing Tang, E-mail: tangfaqing33@hotmail.com; Xiyun Deng, E-mail: dengxiyunmed@hunnu.edu.cn
[1]Key Laboratory of Translational Cancer Stem Cell Research, Hunan Normal University, Changsha, Hunan, China.
[2]Departments of Pathology and Pathophysiology, Hunan Normal University School of Medicine, Changsha, Hunan, China.
[3]Hunan Cancer Hospital & The Affiliated Cancer Hospital of Xiangya School of Medicine, Central South University, Changsha, Hunan, China.

Triple-negative breast cancer (TNBC) generally has a poor prognosis, with high rates of systemic recurrence and refractoriness to conventional therapy regardless of the choice of adjuvant treatment. Conventional chemotherapy remains the primary treatment option for patients with early and advanced-stage TNBC [1], with anthracycline and taxane-based chemotherapy being the mainstay of the therapy. Nevertheless, despite comprehensive and aggressive management, over 50% of TNBC patients (stages I–III) recur, and more than 37% of those patients succumb within 5 years. Immunotherapy is being actively explored in TNBC and clinical trials are showing encouraging results. This chapter will discuss about the advancement in immunotherapy against TNBC focusing on inhibition of several eminent immune checkpoint molecules, which has yielded promising results in both early and advanced-stage TNBC patients.

8.1 Cancer Immunotherapy

The genetic and epigenetic alterations that are characteristic of all cancers provide a diverse set of antigens that the immune system can use to distinguish tumor cells from their normal counterparts [2]. The evasion of immune activation is recognized as a hallmark of cancer and can involve down-regulation of tumor-specific antigens and the major histocompatibility complex (MHC) on the surface of tumor cells [3], chronic activation of humoral immunity, infiltration by cytotoxic T lymphocytes (CTLs) and other inflammatory cells, etc. Together, these mechanisms act in concert to modulate anti-tumor responses and impose an enormous impact on tumor development and progression.

The immune system not only plays a role in tumor initiation and progression, but also participates in recognition and destruction of cancer cells. The anti-tumor immune response dampens tumor development through tumor-directed immune responses. T cells have been the major focus of therapeutic efforts to manipulate endogenous anti-tumor immunity. Two subtypes of T cells are involved in anti-tumor immunity: the CD8+ effector T cells (i.e., CTLs), which recognize and kill antigen-expressing cells; and the CD4+ helper T cells, which orchestrate diverse immune responses [4]. In the case of CTLs, the generation and maintenance of immune responses are controlled by both co-stimulatory and co-inhibitory signaling in addition to antigen recognition by the T cell receptor (TCR) [5].

It should be noted that CTL-mediated cellular immunity plays a major role in anti-tumor immune response, and the ultimate amplitude and quality of the immune response is regulated by a balance between co-stimulatory and inhibitory signals (i.e., immune checkpoints) on CTLs [6]. Both agonists of co-stimulatory receptors and antagonists of inhibitory signals result in the amplification of antigen-specific T cell responses, and are the primary agents in clinical development [7]. However, more efforts are currently being placed on antagonism of the inhibitory signals, which is among the most promising approaches to activating therapeutic anti-tumor immunity.

8.2 Immune Checkpoints as an Important Target of Anti-cancer Therapy

Immune checkpoints refer to a plethora of inhibitory mechanisms hardwired into the immune system that are crucial for maintaining the self-tolerance and modulating the level and duration of immune responses. It is now clear that tumors co-opt certain immune checkpoint pathways as a major mechanism of immune resistance, particularly against T cell-mediated immunity [7]. Tumor cells display dysregulated expression of immune checkpoint proteins as an important mechanism of immune escape.

Many of the immune checkpoints are initiated by ligand-receptor interactions (**Figure 8-1**). Therefore, they can be readily blocked by

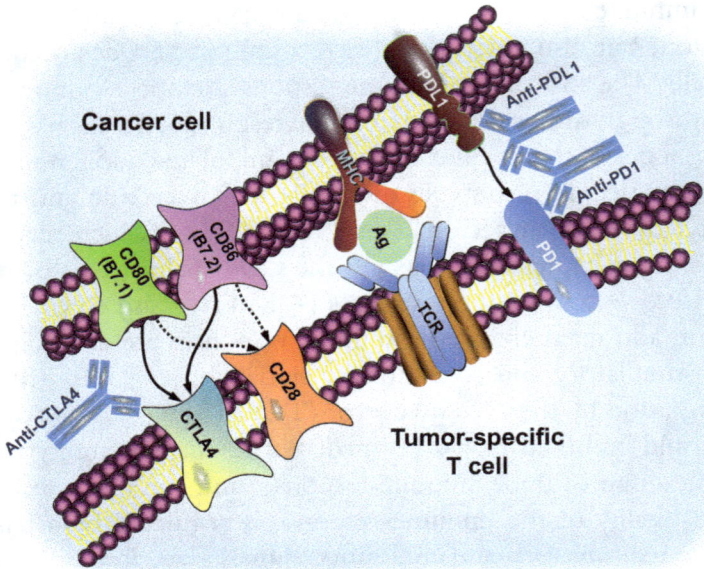

Figure 8-1. Schematic diagram of immune checkpoint blockade. MHC generally presents antigen on the surface of the cancer cells for recognition by CD8⁺ cytotoxic T lymphocyte (CTL) via the TCR. The binding of PD1 on the surface of the CTL with its ligand PDL1 functions to suppress signals downstream of TCR activation, leading to apoptosis of the CTL. CTLA4 is homologous to the T cell co-stimulatory protein CD28, both of which bind to CD80 and CD86 on the surface of cancer cell but with different affinity. Overall, CTLA4 has a much higher affinity than CD28 to CD80/CD86. Antibodies (anti-CTLA4, anti-PD1, anti-PDL1) inhibit these checkpoint-targeting proteins to restore the activity of CTLs and kill cancer cells.

antibodies against the ligands or receptors [7]. Current research is mainly focused on two pairs of inhibitory signals, i.e., the cytotoxic T lymphocyte-associated antigen 4 (CTLA4)/CD80 and PD1/PDL1 receptor/ligand pair [8]. Notably, these immune checkpoint molecules are capable of suppressing T cell responses and are proven to be effective targets in cancer treatment. Blockade of immune checkpoints can reinvigorate dysfunctional/exhausted T cells by restoring immunity to eliminate cancer cells [9]. A number of studies have demonstrated the clinical efficacy of immunotherapy against various types of cancer such as melanomas, leukemia, prostate cancer and lymphomas. The immune checkpoint blocking antibodies have

Figure 8-2. James P. Allison (left) and Tasuku Honjo (right). The Nobel Prize in Physiology or Medicine 2018 was awarded to these two scientists (pictured) for their seminal work on cancer immunotherapy targeting the immune checkpoints.

moved immunotherapy into a new era by targeting the PD1/PDL1 and CTLA4/B7 pathways [10].

In 2018, James P. Allison at the University of Texas MD Anderson Cancer Center and Tasuku Honjo at Kyoto University (pictured) jointly won the Nobel Prize in Physiology or Medicine for their seminal work on CTLA4 and PD1 and designing the strategies for activating the immune system in cancer therapy (**Figure 8-2**).

8.3 Promise of Immunotherapy in TNBC

While TNBC lacks any of the targets for existing agents, and because of its molecular heterogeneity, patients with TNBC respond very different to conventional treatments. Notably, breast cancers of this subtype are found to be immunogenic, and this renders them good candidates for immunotherapy. In recent years, exploiting intrinsic mechanisms of the host immune system to eradicate cancer cells has achieved impressive success, and the advances in immunotherapy have yielded potential new therapeutic strategies for the treatment of this devastating subtype of breast cancer. It is anticipated that the responses initiated by immunotherapeutic interventions will explicitly target and annihilate tumor cells, while at the same time sparing normal cells [11].

The increasing knowledge of cellular immunology and tumor-host immune interactions has led to exciting developments of new immunotherapeutic approaches, including the blockade of immune checkpoints, induction of activation of CTLs [12], adoptive cell transfer (ACT)-based therapy [13], tumor vaccination [14], and modulation of the tumor microenvironment to facilitate CTL activity [2].

For TNBC which has not seen substantial advances in clinical management in recent decades, immunotherapy offers the opportunity for durable response. Unlike other cancer types that respond well to immunotherapy such as lung cancer, melanoma, and bladder cancer, most breast cancers are not inherently immunogenic and typically have a low level of T cell infiltration. However, among the breast cancer subtypes, TNBC has greater tumor immune infiltrate, characterized by higher amount of tumor-infiltrating lymphocytes (TILs), a predictive marker for response to immunotherapy and more favorable survival outcomes [3, 15]. Indeed, increased TILs at diagnosis have been significantly associated with pathologic complete responses (pCR) in breast cancer [16]. Interestingly, the improved overall survival (OS) related to increased TILs is only seen in TNBC and HER2-positive breast cancer. Therefore, manipulation of the immune system represents an attractive strategy for the treatment of TNBC.

In considering immune checkpoint-based immunotherapy, however, identification of the patient groups who would benefit from immunotherapy is a major challenge. Particularly, important consideration should be given to the different subtypes of TNBC (as discussed in Chapter 2), with immunomodulatory (IM) and basal-like (BL) subtypes possessing elevated infiltration of immune cells, and hence, more likely to be responsive to immunotherapy [17]. BL TNBC is known to have high frequency of BRCA1/2 mutations and be genetically unstable, which is another predictor of immunotherapy response [18, 19].

Using four publicly available TNBC genomics datasets, He *et al.* classified TNBC into three classes, namely, Immunity_H, Immunity_M, and Immunity_L [20]. Immunity_H was characterized by greater immune cell infiltration and anti-tumor immune activities, and importantly, better survival prognosis compared to other subtypes. Some cancer-associated pathways were hyperactivated in Immunity_H, including apoptosis, calcium signaling, MAPK signaling, PI3K/Akt signaling, and Ras signaling. In contrast, Immunity_L presented low or depressed immune signatures and increased activation of cell cycle, Hippo signaling, DNA replication and repair, cell adhesion molecule binding, etc.

Recently, TNBC was further classified according to the landscape of tumor microenvironment by scientists at Fudan University Shanghai Cancer Center (FUSCC) using the original multi-omics dataset of TNBC [21]. In a similar way as described above, the TNBC microenvironment phenotypes in this study were classified into three heterogeneous clusters: the "immune-desert", the "innate immune-inactivated", and the "immune-inflamed" cluster. The "immune-desert" cluster was characterized by low microenvironment cell infiltration, which was correlated with Myc amplification. The "innate immune-inactivated" cluster was characterized by infiltration of resting innate immune cells and non-immune stromal cells. In this cluster, chemotaxis but inactivation of innate immunity and low tumor antigen burden might contribute to immune escape, and mutations in the PI3K/Akt pathway might be correlated with this effect. The "immune-inflamed" cluster, with abundant adaptive and innate immune cell infiltration, featured high expression of immune checkpoint molecules. Therefore, stratification of TNBC based on quantitative TIL evaluation, which distinguishes between immune "hot" (high-TIL) and "cold" (low-TIL) tumors, appears to correlate with response to immune checkpoint inhibition and has great clinical implications. Immune checkpoint inhibitors might be effective for the immune "hot" tumors, i.e., the "immune-inflamed" tumors. Transformation of immune "cold" tumors into immune "hot" tumors should be considered for "immune-desert" and "innate immune-inactivated" clusters in order to ensure the efficiency of immunotherapy for TNBC.

8.4 Targeting the PD1/PDL1 Pathway in TNBC

Programmed cell death-1 (PD1, also known as CD279), is emerging as a promising target of cancer immunotherapy [22]. The major role of PD1 is to limit the activity of T cells in peripheral tissues at the time of an inflammatory response to infection and to limit autoimmunity [23]. This translates into a predominant immune resistance mechanism within the tumor microenvironment. PD1 is expressed on the surface of T lymphocytes, which is induced when T cells become activated. The binding of PD1 on T cells by programmed death-ligand 1 (PDL1, also known as B7-H1 or CD274) usually found on cancer cells functions to suppress signals downstream of TCR activation (inducing apoptosis of T cells) [24, 25]. Immunotherapy has shown a great promise in recent studies of breast cancer, and several clinical investigations have suggested that immunotherapeutic approaches have great potential in improving clinical outcomes for patients with this neoplasm [26]. Expression of PD1/PDL1 is associated with higher histologic grades, larger tumor sizes, and triple-negative status, all of which are independent indicators of poor prognosis in breast cancer [27–29].

With a lower side effect profile compared with the anti-CTLA4 strategy, agents targeting PD1 or PDL1 have become a focus of moving immune checkpoint blockade into many tumor types including TNBC. Antibodies against PD1/PDL1 have been approved for the treatment of melanoma, non-small cell lung cancer, Hodgkin's lymphoma, bladder cancer, gastroesophageal cancer, cervical cancer, etc. These include two antibodies against PD1, i.e., pembrolizumab (Keytruda, a humanized monoclonal antibody) and nivolumab (Opdivo, a humanized monoclonal antibody), and three antibodies against PDL1, i.e., atezolizumab (Tecentriq, a humanized monoclonal antibody), avelumab (Bavencio, a humanized monoclonal antibody), and durvalumab (Imfinzi, a humanized monoclonal antibody) (**Table 8-1**).

Particular interest in the clinical management of TNBC would be the combination of immunotherapy with radiotherapy or chemotherapy. Theoretically and practically, these combinations should increase mutational load of tumors, optimize the microenvironment, thus priming the tumor for immunotherapy and improving progression-free survival.

Table 8-1. Summary of immune checkpoint targeting antibodies.

Antibody	Trade name	Target	Category	Approval time	Cancer types
Pembrolizumab	Keytruda	PD1	IgG4	Sep 05, 2014	Unresectable or metastatic melanoma
Nivolumab	Opdivo	PD1	IgG4	Jun 22, 2015	Metastatic non-small cell lung cancer
Atezolizumab	Tecentriq	PDL1	IgG1	May 18, 2016	Locally advanced or metastatic urothelial carcinoma
Avelumab	Bavencio	PDL1	IgG1	Mar 23, 2017	Metastatic Merkel cell carcinoma
Durvalumab	Imfinzi	PDL1	IgG1	May 1, 2017	Locally advanced or metastatic urothelial carcinoma
Ipilimumab	Yervoy	CTLA4	IgG1	Mar 25, 2011	Advanced melanoma
Tremelimumab	\	CTLA4	IgG2	Apr 15, 2015	Malignant mesothelioma

8.4.1 *Humanized Anti-PD1 Antibodies*

8.4.1.1 *Pembrolizumab*

Receiving initial FDA approval for unresectable or metastatic melanoma in 2014, pembrolizumab is one of the best studied immune checkpoint inhibitory agents [30]. In 2016, the KEYNOTE-012 trial, Phase Ib study reported the efficacy with acceptable safety profile when pembrolizumab was given to patients with heavily pre-treated, advanced TNBC [26]. Among the 27 patients evaluated, the overall response rate was 18.5%, with median response time of 17.9 weeks. Further clinical trials were being carried out on metastatic TNBC patients.

Recently, the combination of immunotherapy and chemotherapy had yielded satisfactory results in lung cancer patients [31, 32]. The combined immunotherapy was also being investigated in breast cancer. In locally advanced breast cancer, the addition of pembrolizumab to standard neoadjuvant chemotherapy (paclitaxel followed by doxorubicin and cyclophosphamide) increased the rate of pCR approximately three-fold (60% vs. 20%) [33]. A study presented at American

Society of Clinical Oncology (ASCO) in 2017 reported an objective response rate of 100% with the combination of pembrolizumab and carboplatin vs. 80% in another group (nab-paclitaxel plus pembrolizumab) for BRCA1-mutated breast cancer [18].

The combination of immunotherapy and PARP inhibitors also produced exciting results. The TOPACIO (KEYNOTE-162) clinical trial (NCT02657889) examined the combination of pembrolizumab with niraparib in patients with metastatic TNBC or ovarian cancer. Epacadostat (INCB24360), an oral hydroxyamidine, is an inhibitor of indoleamine-2, 3-dioxygenase 1 (IDO1), which is a tryptophan catabolizing enzyme that induces immune tolerance through T-cell suppression. By inhibiting IDO1 and decreasing kynurenine levels in tumor cells, epacadostat increases and restores the proliferation of immune cells including dendritic cells, NK cells, and T lymphocytes, and decreases Treg cells [34]. The combination of epacadostat with pembrolizumab is currently being tested in a phase I/II study in selected types of cancers including TNBC (KEYNOTE-037/ECHO-202, NCT02178722). Other phase I clinical trials are investigating the combinations of pembrolizumab with kinase inhibitors such as itacitinib, ruxolitinib, binimetinib, or dinaciclib. These studies will determine whether combining targeted therapies with immune-modulatory approaches to annihilate cancer cells is an effective strategy to treat TNBC.

A similar strategy of the combinational use of pembrolizumab with PARP inhibitors yielded an objective response rate of 45% compared to 16.7% in single-agent PARP inhibitor group [35]. Additionally, a clinical trial (NCT03184558) was initiated by BerGenBio ASA to investigate the objective response rate of combination of bemcentinib (BGB324) with pembrolizumab. In this trail, bemcentinib capsules were administered at 400 mg on days 1–3, and 200 mg thereafter, and pembrolizumab was administered at a dose of 200 mg every 3 weeks. Although the study has been completed at August 30, 2018, the results have not been released yet.

8.4.1.2 *Nivolumab*

Nivolumab is another humanized anti-PD1 monoclonal antibody. Due to its significant clinical efficacy against several types of malignancies, nivolumab has become one of the most eye-catching checkpoint inhibitors. A clinical trial (NCT02834247) investigated TAK-659, a selective inhibitor of the Syk tyrosine kinase, in combination with nivolumab in patients with metastatic TNBC. The maximum tolerated dose and the overall response rate were determined after the patients received TAK-659 at 60 mg/day in combination with nivolumab at 3 mg/kg. This study has been finished on November 30, 2018, pending release of the research findings. The TONIC trial is a currently ongoing phase II trial for patients with metastatic TNBC who received ≤3 lines of palliative chemotherapy, and has progressed on the last line of treatment. This is the first trial that has shown promising results using Nivolumab after giving either radiation or chemotherapy. This study is estimated to be completed in August, 2022 [36].

8.4.2 Humanized Anti-PDL1 Antibodies

8.4.2.1 Atezolizumab

Initial phase I findings of atezolizumab, the anti-PDL1 antibody, in metastatic TNBC were recently reported. Of the 9 patients with advanced TNBC evaluated for efficacy, the overall response rate was 33% [37]. Importantly, the PDL1 expression level is predictive of the clinical response: 44.4% in TNBC patients with PDL1-positive immune cells vs. 2.6% in those without PDL1-positive cells.

Recently, a phase III clinical trial evaluating the effects of atezolizumab in combination with nanoparticle albumin-bound (nab)-paclitaxel as first-line treatment in TNBC patients yielded exciting results. Among patients with PDL1-positive tumors, atezolizumab plus nab-paclitaxel significantly prolonged the median OS compared with placebo plus nab-paclitaxel (25.0 vs. 15.5 months) [38]. It should be noted that the benefit is likely to be observed in patients with PDL1-expressing tumors, which account for only a proportion of all TNBC cases. Also, there is a report showing that the glycosylation status of PDL1 has a major impact on the effect of anti-PDL1 immune therapy [39]. Therefore, the selection of TNBC patients that are likely to benefit from immune checkpoint inhibition for optimal outcome would be of great importance. Nonetheless, these findings are expected to lead to new treatment options for patients with TNBC.

8.4.2.2 Avelumab

The JAVELIN phase I study investigated the anti-PDL1 antibody, avelumab, in several types of tumors including 168 breast cancer patients who were not selected for PDL1 expression. The response rate in the whole breast cancer cohort was 4.8%; however, in the TNBC cohort, when patients were selected based on PDL1 expression using cutoff for positivity of ≥10%, the response attained an impressive rate of 44.4%. Two other phase I studies also reported encouraging response rates with checkpoint inhibitors for TNBC patients who were selected for PDL1 positivity [40].

A larger-scale phase Ib study involving 168 patients with 57 being TNBC, who previously received taxane and anthracycline-based therapies, was used to evaluate the efficiency of avelumab. Across the entire cohort, the objective response rate was 5.4% [41]. Recently, avelumab as adjuvant treatment for TNBC was investigated in a phase III randomized trial (NCT02926196), which enrolled 335 patients in. The primary endpoint of this trial is to determine whether adjuvant avelumab (10 mg/kg IV every 2 weeks for 1 year) improves disease-free survival (DFS). The secondary outcomes include OR and safety profile. This study is currently recruiting patients and would be completed in June, 2023.

8.4.2.3 Durvalumab

Several trials were also being performed with durvalumab for patients with metastatic TNBC in combination therapy [42]. A randomized phase II study showed that the addition of durvalumab to an anthracycline taxane-based neoadjuvant therapy in early TNBC significantly increases pCR rate [43].

8.5 Targeting the CTLA4 Molecule in TNBC

CTLA4, also known as CD152, the first co-inhibitory molecule identified and the first immune checkpoint receptor clinically targeted [44], is exclusively expressed on T cells where it primarily regulates the amplitude of early-stage T cell activation. The ligands of CTLA4, i.e., CD80 (also known as B7.1) and CD86 (also known as B7.2), could also bind to the co-stimulatory receptor CD28. Compared with CD28 [45], CTLA4 has a much higher overall affinity for both CD80 and CD86 [46]. Therefore, the expression of CTLA4 on the surface of T cells dampens the activation of T cells by outcompeting CD28's positive co-stimulatory signal. Blocking CTLA4 can reactivate T cells and enhance anti-tumor immune response [47]. This dominance of negative signals from CTLA4-CD80/CD86 interaction results in reduced T cell proliferation and decreased IL-2 production [48]. The central role of CTLA4 in inhibiting T cell activity is demonstrated by the lethal systemic immune hyperactivation phenotype of CTLA4-knockout mice [49]. Significantly, recent phase I and phase II clinical trials confirm the efficacy of anti-CTLA4 strategy for the treatment of cancers. Meanwhile, these studies also indicated an increase in efficacy when combined with PD1 or inhibitors of the Ras/Raf/MAPK/MEK/ERK pathway [47].

8.5.1 *Preclinical Studies of CTLA4 Blockade*

As an important strategy of cancer immunotherapy, CTLA4 blockade results in a broad enhancement of immune responses that are dependent on helper T cells [50]. Initially, the general strategy of blocking CTLA4 was questioned because there is no tumor specificity of the expression of the CTLA4 ligands (other than some myeloid and lymphoid tumors) and because the dramatic lethal autoimmune and hyperimmune phenotype of CTLA4-knockout mice predicted a high degree of immune toxicity associated with blockade of this receptor. However, Allison and colleagues used preclinical models to demonstrate that a therapeutic window was indeed achieved when CTLA4 was partially blocked with antibodies against CTLA4 [51]. The initial studies demonstrated significant anti-tumor responses without overt immune toxicities when the mice bearing partially immunogenic tumors were treated with CTLA4 antibodies as single agents. Poorly immunogenic tumors did not respond to anti-CTLA4 as a single agent but did respond when anti-CTLA4 was combined with a granulocyte-macrophage colony-stimulating factor (GM-CSF)-transduced cellular vaccine [52]. In addition, Liu *et al.* has found that combination immunotherapy of MUC1 (a heavily glycosylated type 1 transmembrane mucin), mRNA nano-vaccine, and anti-CTLA4 monoclonal antibody could significantly enhance anti-tumor immune response to inhibit growth of TNBC [53].

8.5.2 *Humanized Anti-CTLA4 Antibodies*

The above preclinical findings encouraged the development and testing of two humanized CTLA4 antibodies, ipilimumab (trade name Yervoy, also known as MDX-010, MDX-101, or BMS-734016) and tremelimumab (formerly known as ticilimumab or CP-675206) (**Table 8-1**), which began in 2000. As with virtually all anti-cancer agents, initial testing was as a single agent in patients with advanced melanoma and ovarian cancer who did not respond to conventional therapy [54]. Both antibodies produced objective clinical responses in ~10% of patients with melanoma, but immune-related toxicities involving various tissue sites were observed in 25–30% of patients, with colitis being a particularly common event. The first randomized phase III clinical trial was for tremelimumab in patients with advanced melanoma. In this trial, 15 mg per kg tremelimumab was given every three months as a single agent, comparing with dacarbazine, a standard melanoma chemotherapy treatment agent. The trial showed no survival benefit with this dose [55]. Importantly, the stage IV lung metastasis TNBC patients received low-dose immune checkpoint blockade (concurrent nivolumab and ipilimumab) weekly over 3 weeks with regional hyperthermia 3 times a week, followed by systemic fever-range hyperthermia induced by interleukin-2 for 5 day, which might lead to complete remission of pulmonary metastases [56]. Currently, anti-CTLA4 immunotherapy is being tested in non-small cell lung cancer, breast cancer, and melanoma, with a focus on brain metastases, either as monotherapy or in combination with other therapeutic agents [57, 58]. Additionally, DZ-2384 is a candidate microtubule-targeting agent, which is highly effective in patient-derived taxane-sensitive and taxane-resistant xenograft models of TNBC at lower doses and over a wider range relative to paclitaxel. The preclinical trials have demonstrated that DZ-2384 synergistically acts with anti-CTLA4 immunotherapy, which is a promising therapeutic agent for the treatment of metastatic TNBC [59].

8.5.3 *Clinical Trials of Anti-CTLA4 Antibody in TNBC*

The presently available information regarding the clinical use of anti-CTLA4 antibodies in TNBC is limited. One clinical trial (NCT02527434, phase II) examined the use of tremelimumab in combination with another immune checkpoint inhibitor for the treatment of advanced solid tumors including TNBC, pending release of the trial results. Additionally, another clinical phase II (NCT03818685) trial is evaluating the clinical benefit of a post-operative adjuvant therapy comparing between radiotherapy plus nivolumab and ipilimumab vs. radiotherapy plus capecitabine in TNBC patients with residual disease after neoadjuvant chemotherapy.

References

1. Oualla K, El-Zawahry HM, Arun B, Reuben JM, Woodward WA, Gamal El-Din H, Lim B, Mellas N, Ueno NT, Fouad TM. (2017). Novel therapeutic strategies in the treatment of triple-negative breast cancer. *Ther Adv Med Oncol*, 9(7): 493–511.
2. Qian X, Zhang Q, Shao N, Shan Z, Cheang T, Zhang Z, Su Q, Wang S, Lin Y. (2019). Respiratory hyperoxia reverses immunosuppression by regulating myeloid-derived suppressor cells and PD-L1 expression in a triple-negative breast cancer mouse model. *Am J Cancer Res*, 9(3): 529–545.
3. Vikas P, Borcherding N, Zhang W. (2018). The clinical promise of immunotherapy in triple-negative breast cancer. *Cancer Manag Res*, 10: 6823–6833.
4. Lin CT, Chang TC, Shaw SW, Cheng PJ, Huang CT, Chao A, Soong YK, Lai CH. (2006). Maintenance of CD8 effector T cells by CD4 helper T cells eradicates growing tumors and promotes long-term tumor immunity. *Vaccine*, 24(37–39): 6199–6207.

5. Peggs KS, Quezada SA, Allison JP. (2009). Cancer immunotherapy: Co-stimulatory agonists and co-inhibitory antagonists. *Clin Exp Immunol*, **157**(1): 9–19.

6. Zou W, Chen L. (2008). Inhibitory B7-family molecules in the tumour microenvironment. *Nat Rev Immunol*, **8**(6): 467–477.

7. Pardoll DM. (2012). The blockade of immune checkpoints in cancer immunotherapy. *Nat Rev Cancer*, **12**(4): 252–264.

8. Gudi RR, Karumuthil-Melethil S, Perez N, Li G, Vasu C. (2019). Engineered dendritic cell-directed concurrent activation of multiple T cell inhibitory pathways induces robust immune tolerance. *Sci Rep*, **9**(1): 12065.

9. van der Vlist M, Kuball J, Radstake TR, Meyaard L. (2016). Immune checkpoints and rheumatic diseases: What can cancer immunotherapy teach us? *Nat Rev Rheumatol*, **12**(10): 593–604.

10. Tsai HF, Hsu PN. (2017). Cancer immunotherapy by targeting immune checkpoints: Mechanism of T cell dysfunction in cancer immunity and new therapeutic targets. *J Biomed Sci*, **24**(1): 35–42.

11. Jia H, Truica CI, Wang B, Wang Y, Ren X, Harvey HA, Song J, Yang JM. (2017). Immunotherapy for triple-negative breast cancer: Existing challenges and exciting prospects. *Drug Resist Updat*, **32**: 1–15.

12. Parvizpour S, Razmara J, Pourseif MM, Omidi Y. (2019). In silico design of a triple-negative breast cancer vaccine by targeting cancer testis antigens. *Bioimpacts*, **9**(1): 45–56.

13. Assadipour Y, Zacharakis N, Crystal JS, Prickett TD, Gartner JJ, Somerville RPT, Xu H, Black MA, Jia L, Chinnasamy H *et al.* (2017). Characterization of an immunogenic mutation in a patient with metastatic triple-negative breast cancer. *Clin Cancer Res*, **23**(15): 4347–4353.

14. Chen Z, Hu K, Feng L, Su R, Lai N, Yang Z, Kang S. (2018). Senescent cells re-engineered to express soluble programmed death receptor-1 for inhibiting programmed death receptor-1/programmed death ligand-1 as a vaccination approach against breast cancer. *Cancer Sci*, **109**(6): 1753–1763.

15. Stagg J, Allard B. (2013). Immunotherapeutic approaches in triple-negative breast cancer: Latest research and clinical prospects. *Ther Adv Med Oncol*, **5**(3): 169–181.

16. Adams S, Gray RJ, Demaria S, Goldstein L, Perez EA, Shulman LN, Martino S, Wang M, Jones VE, Saphner TJ *et al.* (2014). Prognostic value of tumor-infiltrating lymphocytes in triple-negative breast cancers

from two phase III randomized adjuvant breast cancer trials: ECOG 2197 and ECOG 1199. *J Clin Oncol*, **32**(27): 2959–2966.

17. Haffty BG. (2009). Locoregional recurrence of triple-negative breast cancer after breast-conserving surgery and radiation. *Cancer*, **20**(3): 297–298.

18. Nolan E, Savas P, Policheni AN, Darcy PK, Vaillant F, Mintoff CP, Dushyanthen S, Mansour M, Pang JB, Fox SB *et al.* (2017). Combined immune checkpoint blockade as a therapeutic strategy for BRCA1-mutated breast cancer. *Sci Transl Med*, **9**(393): eaal4922.

19. Hermsen BB, Verheijen RH, Menko FH, Gille JJ, van Uffelen K, Blankenstein MA, Meijer S, van Diest PJ, Kenemans P, von Mensdorff-Pouilly S. (2007). Humoral immune responses to MUC1 in women with a BRCA1 or BRCA2 mutation. *Eur J Cancer*, **43**(10): 1556–1563.

20. He Y, Jiang Z, Chen C, Wang X. (2018). Classification of triple-negative breast cancers based on Immunogenomic profiling. *J Exp Clin Cancer Res*, **37**(1): 327–339.

21. Xiao Y, Ma D, Zhao S, Suo C, Shi J, Xue MZ, Ruan M, Wang H, Zhao J, Li Q *et al.* (2019). Multi-omics profiling reveals distinct microenvironment characterization and suggests immune escape mechanisms of triple-negative breast cancer. *Clin Cancer Res*, **25**(16): 5002–5014.

22. Homet Moreno B, Ribas A. (2015). Anti-programmed cell death protein-1/ligand-1 therapy in different cancers. *Br J Cancer*, **112**(9): 1421–1427.

23. Sharpe AH, Pauken KE. (2018). The diverse functions of the PD1 inhibitory pathway. *Nat Rev Immunol*, **18**(3): 153–167.

24. Iwai Y, Ishida M, Tanaka Y, Okazaki T, Honjo T, Minato N. (2002). Involvement of PD-L1 on tumor cells in the escape from host immune system and tumor immunotherapy by PD-L1 blockade. *Proc Natl Acad Sci U S A*, **99**(19): 12293–12297.

25. Emens LA. (2018). Breast cancer immunotherapy: Facts and hopes. *Clin Cancer Res*, **24**(3): 511–520.

26. Nanda R, Chow LQ, Dees EC, Berger R, Gupta S, Geva R, Pusztai L, Pathiraja K, Aktan G, Cheng JD *et al.* (2016). Pembrolizumab in patients with advanced triple-negative breast cancer: Phase Ib KEYNOTE-012 study. *J Clin Oncol*, **34**(21): 2460–2467.

27. Ghebeh H, Mohammed S, Al-Omair A, Qattan A, Lehe C, Al-Qudaihi G, Elkum N, Alshabanah M, Bin Amer S, Tulbah A *et al.* (2006). The B7-H1 (PD-L1) T lymphocyte-inhibitory molecule is expressed in breast cancer

patients with infiltrating ductal carcinoma: Correlation with important high-risk prognostic factors. *Neoplasia*, **8**(3): 190–198.

28. Mittendorf EA, Philips AV, Meric-Bernstam F, Qiao N, Wu Y, Harrington S, Su X, Wang Y, Gonzalez-Angulo AM, Akcakanat A *et al.* (2014). PD-L1 expression in triple-negative breast cancer. *Cancer Immunol Res*, **2**(4): 361–370.

29. Muenst S, Soysal SD, Gao F, Obermann EC, Oertli D, Gillanders WE. (2013). The presence of programmed death 1 (PD-1)-positive tumor-infiltrating lymphocytes is associated with poor prognosis in human breast cancer. *Breast Cancer Res Treat*, **139**(3): 667–676.

30. Redman JM, Gibney GT, Atkins MB. (2016). Advances in immuno-therapy for melanoma. *BMC Med*, **14**(1): 20–30.

31. Langer CJ, Gadgeel SM, Borghaei H, Papadimitrakopoulou VA, Patnaik A, Powell SF, Gentzler RD, Martins RG, Stevenson JP, Jalal SI *et al.* (2016). Carboplatin and pemetrexed with or without pembrolizumab for advanced, non-squamous non-small-cell lung cancer: A randomised, phase 2 cohort of the open-label KEYNOTE-021 study. *Lancet Oncol*, **17**(11): 1497–1508.

32. Gandhi L, Rodriguez-Abreu D, Gadgeel S, Esteban E, Felip E, De Angelis F, Domine M, Clingan P, Hochmair MJ, Powell SF *et al.* (2018). Pembrolizumab plus chemotherapy in metastatic non-small-cell lung cancer. *N Engl J Med*, **378**(22): 2078–2092.

33. Geyer FC, Magali LT, Pierre-Emmanuel C, Neill P, Arnaud G, Rachael N, Lambros MBK, Ibrahim K, Constance A, Sandra O. (2012). Molecular evidence in support of the neoplastic and precursor nature of microglandular adenosis. *Histopathology*, **60**(6B): E115–E130.

34. Jochems C, Fantini M, Fernando RI, Kwilas AR, Donahue RN, Lepone LM, Grenga I, Kim YS, Brechbiel MW, Gulley JL *et al.* (2016). The IDO1 selective inhibitor epacadostat enhances dendritic cell immunoge-nicity and lytic ability of tumor antigen-specific T cells. *Oncotarget*, **7**(25): 37762–37772.

35. Geyer FC, Weigelt B, Natrajan R, Lambros MB, de Biase D, Vatcheva R, Savage K, Mackay A, Ashworth A, Reis-Filho JS. (2010). Molecular analysis reveals a genetic basis for the phenotypic diversity of metaplastic breast carcinomas. *J Pathol*, **220**(5): 562–573.

36. Guerini-Rocco E, Hodi Z, Piscuoglio S, Ng CK, Rakha EA, Schultheis AM, Marchia[2] C, Da CPA, De Filippo MR, Martelotto LG. (2015). The repertoire of somatic genetic alterations of acinic cell carcinomas of the breast: An exploratory, hypothesis-generating study. *J Pathol*, **237**(2): 166–178.

37. Yaqub F. (2015). 2014 San Antonio breast cancer symposium. *Lancet Oncol*, **16**(2): 135–136.

38. Schmid P, Adams S, Rugo HS, Schneeweiss A, Barrios CH, Iwata H, Dieras V, Hegg R, Im SA, Shaw Wright G *et al.* (2018). Atezolizumab and nab-paclitaxel in advanced triple-negative breast cancer. *N Engl J Med*, **379**(22): 2108–2121.

39. Li CW, Lim SO, Chung EM, Kim YS, Park AH, Yao J, Cha JH, Xia W, Chan LC, Kim T *et al.* (2018). Eradication of triple-negative breast cancer cells by targeting glycosylated PD-L1. *Cancer Cell*, **33**(2): 187–201 e110.

40. Dirix LY, Takacs I, Jerusalem G, Nikolinakos P, Arkenau HT, Forero-Torres A, Boccia R, Lippman ME, Somer R, Smakal M *et al.* (2018). Avelumab, an anti-PD-L1 antibody, in patients with locally advanced or metastatic breast cancer: A phase 1b JAVELIN Solid Tumor study. *Breast Cancer Res Treat*, **167**(3): 671–686.

41. Hagen K, Rinat Y, Ryan W, Cheang MCU, David V, Speers CH, Nielsen TO, Karen G. (2010). Metastatic behavior of breast cancer subtypes. *J Clin Oncol*, **28**(20): 3271–3277.

42. Katz H, Alsharedi M. (2017). Immunotherapy in triple-negative breast cancer. *Med Oncol*, **35**(1): 13.

43. Loibl S, Untch M, Burchardi N, Huober J, Sinn BV, Blohmer JU, Grischke EM, Furlanetto J, Tesch H, Hanusch C *et al.* (2019). A randomised phase II study investigating durvalumab in addition to an anthracycline taxane-based neoadjuvant therapy in early triple-negative breast cancer: Clinical results and biomarker analysis of GeparNuevo study. *Ann Oncol*, **30**(8): 1279–1288.

44. Lo B, Abdel-Motal UM. (2017). Lessons from CTLA-4 deficiency and checkpoint inhibition. *Curr Opin Immunol*, **49**: 14–19.

45. Krummel MF, Allison JP. (1995). CD28 and CTLA-4 have opposing effects on the response of T cells to stimulation. *J Exp Med*, **182**(2): 459–465.

46. Walunas TL, Bakker CY, Bluestone JA. (1996). CTLA-4 ligation blocks CD28-dependent T cell activation. *J Exp Med*, **183**(6): 2541–2550.

47. Szostak B, Machaj F, Rosik J, Pawlik A. (2019). CTLA4 antagonists in phase I and phase II clinical trials, current status and future perspectives for cancer therapy. *Expert Opin Investig Drugs*, **28**(2): 149–159.

48. Rudd CE, Taylor A, Schneider H. (2009). CD28 and CTLA-4 coreceptor expression and signal transduction. *Immunol Rev*, **229**(1): 12–26.

49. Waterhouse P, Penninger JM, Timms E, Wakeham A, Shahinian A, Lee KP, Thompson CB, Griesser H, Mak TW. (1995). Lymphoproliferative disorders with early lethality in mice deficient in Ctla-4. *Science*, **270**(5238): 985–988.

50. Chambers CA, Kuhns MS, Egen JG, Allison JP. (2001). CTLA-4-mediated inhibition in regulation of T cell responses: Mechanisms and manipulation in tumor immunotherapy. *Annu Rev Immunol*, **19**: 565–594.

51. Leach DR, Krummel MF, Allison JP. (1996). Enhancement of antitumor immunity by CTLA-4 blockade. *Science*, **271**(5256): 1734–1736.

52. Chen P, Chen F, Zhou B. (2018). Comparisons of therapeutic efficacy and safety of ipilimumab plus GM-CSF versus ipilimumab alone in patients with cancer: A meta-analysis of outcomes. *Drug Des Devel Ther*, **12**: 2025–2038.

53. Liu L, Wang Y, Miao L, Liu Q, Musetti S, Li J, Huang L. (2018). Combination immunotherapy of MUC1 mRNA nano-vaccine and CTLA-4 blockade effectively inhibits growth of triple negative breast cancer. *Mol Ther*, **26**(1): 45–55.

54. Hodi FS, Mihm MC, Soiffer RJ, Haluska FG, Butler M, Seiden MV, Davis T, Henry-Spires R, MacRae S, Willman A *et al.* (2003). Biologic activity of cytotoxic T lymphocyte-associated antigen 4 antibody blockade in previously vaccinated metastatic melanoma and ovarian carcinoma patients. *Proc Natl Acad Sci USA*, **100**(8): 4712–4717.

55. Ribas A. (2010). Clinical development of the anti-CTLA-4 antibody tremelimumab. *Semin Oncol*, **37**(5): 450–454.

56. Kleef R, Moss R, Szasz AM, Bohdjalian A, Bojar H, Bakacs T. (2018). Complete clinical remission of stage IV triple-negative breast cancer lung metastasis administering low-dose immune checkpoint blockade in combination with hyperthermia and Interleukin-2. *Integr Cancer Ther*, **17**(4): 1297–1303.

57. Blank CU, Enk A. (2015). Therapeutic use of anti-CTLA-4 antibodies. *Int Immunol*, **27**(1): 3–10.

58. Venur VA, Ahluwalia MS. (2017). Novel therapeutic agents in the management of brain metastases. *Curr Opin Oncol*, **29**(5): 395–399.

59. Bernier C, Soliman A, Gravel M, Dankner M, Savage P, Petrecca K, Park M, Siegel PM, Shore GC, Roulston A. (2018). DZ-2384 has a superior preclinical profile to taxanes for the treatment of triple-negative breast cancer and is synergistic with anti-CTLA-4 immunotherapy. *Anticancer Drugs*, **29**(8): 774–785.

Appendix: Online Resources

1. Triple-Negative Breast Cancer: Overview, Treatment, and More
   https://www.breastcancer.org/symptoms/diagnosis/trip_neg
   This website contains news, informational booklets, online discussion boards, blog, chat rooms, and ask-the-expert conferences regarding TNBC.
2. Susan G. Komen: Triple Negative Breast Cancer Facts
   https://ww5.komen.org/breastcancer/triplenegativebreastcancer.html
   This website contains information on TNBC, including fact sheets and survivor stories.
3. Living Beyond Breast Cancer: Guide to Understanding TNBC – 5th Edition
   https://www.lbbc.org/get-support/print/guides-to-understanding/guide-understanding-triple-negative-breast-cancer
   This brochure was created in partnership with Triple-Negative Breast Cancer Foundation and contains information for those diagnosed with TNBC, including facts regarding potential treatments and post-treatment concerns. Experiences of real women affected by TNBC and tips from professionals are included.
4. Triple Negative Breast Cancer Foundation
   www.tnbcfoundation.org
   This website contains information for patients, scientists, and patient advocacy groups.

5. State-of-the-art Treatment for TNBC: Talking with the Experts
   https://tnbc-prd.s3.amazonaws.com/8b74dd9e-b689-4bc6-afd9-e0942752f89a/State-of-the-Art-Treatment-for-TNBC.pdf
   This website contains information for patients, scientists, and patient advocacy groups.

6. National Comprehensive Cancer Network: Guidelines for Patients with Breast Cancer
   http://www.nccn.org/patients/guidelines/cancers.aspx#breast
   This website contains guidelines to help patients better understand their treatment and increase their participation in conversations with healthcare providers.

7. National Institutes of Health: ClinicalTrials.gov
   www.clinicaltrials.gov
   This is a registry and database of publicly and privately supported clinical studies of human participants conducted around the world.

8. National Cancer Institute: Clinical Trials Database
   www.cancer.gov/clinicaltrials
   This website contains information on clinical trials, including trials accepting patients with breast cancer.

Afterword

Upon completion of the full manuscript, I owe a lot to the authors of each chapter for their time and effort in manuscript writing and many rounds of revisions. I am especially grateful to the members of my work team and my family. Without the passionate and spirited support from all of you, the publication of this book would not have been possible. We are also greatly indebted and apologetic to those investigators whose important works are not cited in this edition. We will try to include all the important developments in the field of TNBC research in our next edition. Owing to the limitations of the level of our knowledge and understanding, mistakes and flaws are unavoidable — of which we are completely aware. Your critical comments and suggestions are warmly welcome and will be greatly appreciated. Finally, our special thanks ought to be given to the editors and production team at World Scientific Publishing for their professional help and hard work to allow the timely publication of the book. A medical professional or a patient — may the love and blessings of Almighty God be with you, always!

Xiyun Deng, Ph.D.
Director, Key Laboratory of Translational Cancer Stem Cell Research
Hunan Normal University
China

Index